T0257121

Zwicky

Zwicky

THE OUTCAST GENIUS
WHO UNMASKED THE UNIVERSE

JOHN JOHNSON JR.

Harvard University Press

Cambridge, Massachusetts
London, England
2019

Copyright © 2019 by John Johnson Jr.
All rights reserved
Printed in the United States of America

First printing

Library of Congress Cataloging-in-Publication Data
Names: Johnson, John, 1947– author.
Title: Zwicky : the outcast genius who unmasked the universe / John Johnson, Jr.
Description: Cambridge, Massachusetts : Harvard University Press, 2019. |
Includes bibliographical references and index.
Identifiers: LCCN 2019009091 | ISBN 9780674979673 (alk. paper)
Subjects: LCSH: Zwicky, F. (Fritz), 1898–1974. |
Astrophysics—History—20th century. | Science news.
Classification: LCC QB460.72.Z35 J64 2019 | DDC 523.01092 [B]—
dc23 LC record available at https://lccn.loc.gov/2019009091

For Dylan

More than a son, a friend

CONTENTS

Zwicky

⁎ 1 ⁎

NOTED YOUNG MEN
OF SCIENCE

IN THE LATE NINETEENTH century, America was fast becoming an industrial colossus. The last spike completing the transcontinental railroad was driven home. Andrew Carnegie's mighty mills in Pennsylvania spat fire, and skyscrapers rose in New York. Innovations in transportation—the automobile and the airplane—lay just around the corner.

In one area, however, the United States remained decidedly provincial: the physical sciences. Americans proud of their nation-building were startled by a sudden flow of discoveries from Europe describing invisible forces and unseen particles that zipped around at dizzying speeds. In 1895, while working in his lab in Germany, Wilhelm Roentgen noticed a strange glow coming from a glass tube in which he had installed electrodes. His accidental discovery of X-rays set the stage for the detection of radioactivity. Then came the electron, the dancing, shapeless particle that stands at the gateway to the atom, the power of which would transform and nearly destroy civilization in the next century.

Newspaper readers in America were amazed by these discoveries, even if they didn't quite understand what it all meant. Scientists, on the other hand, considered them disturbing proof of America's slow-footedness when it came to plumbing the new, unseen regions of the universe. The nation's first physics professors were appointed only in the 1870s. These early pioneers often had to use their own funds to equip their laboratories.

But by 1899, there were enough of them to assemble for a meeting in New York, from which emerged the American Physical Society.[1]

The society's founders all came from elite eastern academies. Traditionally, that was where the deepest thinking went on. People only moved out west for their health, to paint Native Americans, or to break into movies. By the 1930s, however, the center of gravity for scientific research in the United States was beginning to shift westward.

In Southern California, Robert Millikan won a Nobel Prize for measuring the charge of the electron. By the third decade of the twentieth century, he was transforming a former teacher's college into a premier physical sciences institution called the California Institute of Technology.

Something similar was happening in Northern California, where the University of California at Berkeley scored a major coup by outbidding Yale for the services of Ernest O. Lawrence, the brilliant South Dakotan who would construct the world's first particle accelerator. Made of glass, sealing wax, and bronze, with a kitchen chair and a clothes tree for support, the Rube Goldberg–like device proved that the best way to pierce the atom's inner sanctum was to whirl particles around and around and send them storming into an atomic nucleus. Lawrence's eleven-inch cyclotron, built in 1931, was the forerunner of Europe's giant, the Large Hadron Collider, seventeen miles in circumference.

Given all this activity on the West Coast, it was only natural that the Physical Society would begin to hold regular meetings there. The site of the December 1933 meeting was Stanford University, an institution that was already building a reputation for excellence that rivaled some of the best eastern institutions. Founded by Leland Stanford, a former governor of California who made his fortune in railroads, the university had been a stock farm. The small town that grew up around it was called Palo Alto, or "tall tree" in Spanish, after a single giant redwood that grew along San Francisquito Creek.

By 1933, the university had assembled a respected faculty, but there was little expectation that anything important would occur at the Physical Society meeting, which began on December 15 in the main lecture room of the Physics Department. Forty papers were presented to an audience of only sixty, an indication of how modest the session was expected to be.[2] Similar meetings in New York at that time drew hundreds of onlookers and physicists. All of the papers were produced by western scientists, the subjects ranging from the physics of crystals to the radium content of lava flows in Lassen Volcanic National Park in Northern California.

The subject that produced the most research was cosmic radiation, now called cosmic rays. It was one of the deeper mysteries to trouble the scientific world at the time. The fact that invisible forces shape our world was still something of a novel concept, so the idea of cosmic radiation flitting around Earth, pushing its way into buildings and penetrating bodies, seemed simply weird. The radiation—which we now know as the emission of energy from subatomic particles, mostly free-range protons—first appeared in electroscopes, one of the first scientific instruments to chart the effects of electricity. An electroscope consists of a metal rod from which are suspended two gold leaves; when the rod is exposed to a source of electricity, the two leaves repel each other because both leaves have the same charge. But after a time, an unaccountable thing happens: the leaves spontaneously discharge. Each leaf falls back to its original position. Scientists realized some form of radiation was penetrating the chamber and robbing the leaves of their charges. But what was it? And where did it come from?

Some believed the strange radiation came from somewhere far off in space. Others thought there had to be a local source, maybe under the sea, maybe in the still uncharted wilderness of the solar system. Caltech's Paul Epstein, a Russian-born immigrant who made important contributions to quantum theory, presented a paper at the conference coming down squarely on the side of locality. He argued that the rays "can travel only a finite distance before completely losing their energy."[3] So they had to be local. As support for this idea, he cited a theory by a young colleague named Fritz Zwicky. Zwicky, who had come to America only a few years earlier, was the brashly brilliant son of a Swiss factory owner. He argued that photons of light lose energy as they travel vast, cosmological distances.

Zwicky proposed his tired light theory to challenge the recent discovery that the universe was expanding, or "blowing up," as the newspaper people put it. Unafraid of assailing majority opinion, Zwicky called those who believed in the expanding universe "horse's asses." That was one of the milder epithets the 35-year-old physicist employed against those who disagreed with him.

Another colleague at Caltech, R. M. Langer, imagined an undefined "cosmical field" as the source of the radiation.[4] That idea went nowhere, but Langer would earn lasting fame later on as one of the most imaginative prophets of the new scientific age of atomic power. He would write that the power of the atom would produce a second Garden of Eden, where people would zip around in atomic cars and atomic planes and energy would be so cheap it would be essentially free. "Privilege and class distinctions and the other sources of social uneasiness and bitterness will become relics because things that make up *the good life* will be so abundant and inexpensive," Langer predicted.[5] The irradiated good life remained a vision unfulfilled when the first use of atomic energy was not to propel cars and planes but to reduce two Japanese cities to rubble.

As it happened, Zwicky himself was in attendance at the 1933 Stanford meeting. With his colleague Walter Baade, he was about to propose a set of theories that would rival Einstein's in their imagination and presumption, transforming the humble Stanford meeting into an unexpected crossroads of scientific history. Zwicky's journey to these theories may not have been as dangerous as the voyage of the *Beagle,* but by revealing the great life and death struggles in the heavens, the result was as important to understanding the evolution of the universe as Darwin's journey was to understanding the progress of life on Earth.

Baade and Zwicky were an odd pairing. Baade, who was on the staff of the Mount Wilson Observatory north of Los Angeles, was polite, respectful, and walked with a pronounced limp from a congenital deformity. Exacting in everything he did, no one was better at manipulating the new astronomical tools to tease out secrets from the cosmos. Zwicky, on the other hand, was bold, imaginative, and instinctual. A roughneck mountain climber who never shied from a good battle, whether it was with nature or a superior at Caltech, he had formed an unlikely but productive alliance with Baade.

Zwicky, always a witty and relaxed public speaker, even if his Swiss-German accent was still very much in evidence, presented their joint paper, titled "Supernovae and Cosmic Rays."[6] Supernovae were a special class of exploding stars. They were not unknown at the time, but they were a curiosity—circus freaks among the stars—rather than essential to understanding the great forces at work in the universe. There is debate

Caltech in 1933: Fritz Zwicky, at left, is in consultation with Niels Bohr. While Bohr was thinking and writing about the landscape of the atom, Zwicky was theorizing about supernovae and neutron stars.

about who introduced the term. Many credit it to Zwicky, but the basic idea appears to have come from the Swedish astronomer Knut Lundmark, who in 1920 referred to a special class of novae—or ordinary exploding stars—that he called "giant novae."[7] Nevertheless, it would be Zwicky who made the term as familiar and exciting to the interested public as black holes would be to later generations.

Zwicky and Baade's idea that supernovae could produce cosmic rays was remarkable in that only a handful of giant exploding stars had been observed, and none by either Zwicky or Baade. The most famous supernova was Tycho's Star, named for the early Danish astronomer, Tycho Brahe. A nobleman, who is often remembered for having lost a chunk of his nose in a duel with a fellow student over who was the better mathematician, Brahe was out walking on the evening of November 11, 1572, when he spotted a "strange star" in the constellation Cassiopeia.[8] Not trusting his own eyes, Brahe quizzed his servants, as well as the occupants of a passing carriage. Only when they said they saw it, too, was

Brahe satisfied that he had indeed witnessed a startling event, the explosion of a star in our own neighborhood of space. At its brightest, Tycho's Star rivaled Venus and remained visible to the naked eye for two years before fading away. Such events were at the time seen as evil omens, evidence that the heavens were disturbed.

According to notes from the Stanford session, preserved in the archives of the Physical Society, Zwicky said such "temporary stars," as they were known at the time, flared up in every stellar system once every few centuries. How he knew this he didn't say. Producing a light show equal to 100 million normal stars, a floodlight among lighted matches, a supernova would outshine the galaxy it was in.[9] The star's explosion, according to Zwicky, would yield hellish temperatures of 2.5 million degrees centigrade. Relying on his and Baade's calculations and his own suppositions, Zwicky suggested that these temporary stars were in the process of self-annihilation, eviscerating themselves down to the atomic level. In "the supernova process (a star's) mass in bulk is annihilated," he said.[10]

All this was just a warm-up for his bold theory about cosmic radiation. Given certain behaviors of cosmic rays, there were three possibilities for their origin, according to Zwicky: that the rays came from empty, intergalactic space, or that they were "survivors from a time when physical conditions in the universe were entirely different from what they are now." Both of those ideas were unsatisfactory, he said. "If, however, the production of cosmic rays is related to some sporadic process, such as the flare-up of a super-nova, the above-mentioned difficulties disappear," Zwicky and Baade said in a paper published several months later.[11]

Zwicky went on to show that the intensities of the rays were in "surprisingly good agreement" with what should be expected if they came from a giant exploding star. Based on the stupendous amount of light such a flare-up produced, each proton of energy must be equal to at least 100 million volts. (Today, we know that GCRs—galactic cosmic rays—can reach energies much greater, as the particles reach 87 percent of the speed of light.) That was why the radiation was so strong when it reached Earth, even after traveling enormous distances. The hypothesis also seemed to answer the question of why no cosmic rays had been detected from the Milky Way. "The reason is simply that no super-nova eruption (had recently) occurred in our galaxy."[12]

All this was new and provocative, but Zwicky was not done. "With all reserve," Zwicky said, with in fact very little reserve, "we advance the view that a super-nova represents the transition of an ordinary star into a neutron star, consisting mainly of neutrons." Such a star would have shrunk far below the size of the original star, with a corresponding increase in density and gravity. The "'gravitational packing' . . . may become very large, and . . . far exceed the ordinary nuclear packing fractions."[13]

According to their theory, after exploding, a supernova would undergo a catastrophic collapse, blowing off most of its electrons and protons while crunching others into a nugget more densely packed than anything ever contemplated. At the time, it was beginning to be understood that there was a lot of empty space in an atom. We now know that more than 99 percent of it is empty; if a hydrogen atom were the size of Earth, the proton at its center would be only around two hundred yards wide. But no one had dared imagine a process that could smash through the atom's borders and go on crunching until the last breath was sucked out of it. A star 500,000 miles across, according to Zwicky, would be reduced to a searing marble only nineteen miles wide. A teaspoonful of the stuff would weigh 10 million pounds.

It's important to remember that the term *black hole* had not yet been invented.[14] In fact, the neutron had only been discovered the year before Zwicky unveiled the new theory. This made the prediction inspired and breathtakingly cheeky, as if Zwicky had described the age of the dinosaurs from a wishbone.

The scientific revolutionaries admitted that this suggestion invoked processes and energies that had never been observed in nature. "We are fully aware that our suggestion carries with it grave implications regarding the ordinary views about the constitution of stars," they concluded.[15]

Zwicky had first unveiled his idea of neutron stars in a lecture at Caltech a month earlier. But describing unproven theories to students eager to imagine wondrous things in the heavens was not the same as announcing heretical ideas to a conference of experts, who thought they had a pretty good grasp of what the universe was like.

The Physical Society did not preserve the reaction from the scientists at Stanford. But in a letter to a Swiss friend, Zwicky described it in amused terms. "With a German astronomer I have cooked up a new theory for cosmic radiation," he said. The prediction of neutron stars was far more important, but no one could anticipate how much. "And if the indignation of the foreign physicists takes on the same dimensions as those of the local ones, then we will have to fear for our lives. When I presented the new theory here in a seminar there was such an uproar that some of the more conservative gentlemen nearly died of heart attacks."[16]

Zwicky's description revealed more than his confidence in his new theories. It showed his preternatural ability to welcome opposition as proof that he was on the right track. It was a characteristic that would underpin all the accomplishments of his working life, one that would bring him both honor and calumny. It lay behind his prediction of dark matter, the scaffolding on which the stars and galaxies are arranged like ornaments on a Christmas tree. And it was critical to his research into jet propulsion and rocket fuels during and after World War II, which helped transform the humble rocket, a toy of backyard dabblers, into ballistic missiles capable of both ending life on Earth and carrying astronauts to the moon. It also contributed to his reputation as a difficult, enigmatic man. Feuding with many of the important scientists of his day, he inspired so much resentment that after his death his critics did all they could to forget or disparage what he had done. Like the great forces he chronicled, Fritz Zwicky distorted the orbits of everyone who came in contact with him, attracting many, driving just as many away.

Zwicky and Baade made several errors in their presentation. At one point Zwicky suggested that before exploding, supernovae could be "quite ordinary stars" like our sun. In fact, only certain kinds of stars can go super, and none is like our sun.[17]

Only giant stars, those with at least eight times the mass of our sun, are candidates for Zwicky's type of supernova, which occurs when the star begins to run out of fuel and its internal furnace can no longer support the star's enormous bulk and gravity. Within one second the giant star collapses, and then the core begins to form a neutron star, just as Zwicky theorized. At that point, when the core reaches temperatures thousands of times hotter than the surface of the sun, a shock wave rebounds outward, causing a powerful explosion and ejecting the outer layers of the star at nearly a tenth of the speed of light.

Zwicky could perhaps be forgiven his mistakes because he depended on old and unreliable data for his research; the most recent supernova for which there were good records occurred in 1885. It was long gone by 1934. No remains could be found, not even with the 100-inch Hooker telescope, the world's most powerful, located atop Mount Wilson north of Los Angeles.

For the purposes of their proposal, Zwicky and Baade assumed that a supernova occurs in a given nebula, or galaxy, once every thousand years. This also would be proven incorrect as time went on. It's now known that, despite their rarity, supernovae occur every fifty years or so in a galaxy such as our own. Scientists today have listed several candidates for a supernova in our portion of the sky, including Betelgeuse and Antares.[18] They could explode at any time in the next thousand to 100 million years, which, in cosmology, is a fairly exact prognostication. Betelgeuse is such a giant, just 640 light years away, that its self-destruction will outshine the moon at night—a fantastic show for those lucky enough to be around.

Observers might not have to wait centuries to see an eruption, the two wrote. Because there were around a thousand galaxies in our region of the universe, odds suggested that "one super-nova per year should be expected in this 'immediate' neighborhood of ours."[19]

This suggestion triggered a worldwide hunt for exploding stars, led, appropriately, by Zwicky himself.

In time, the presentation at Stanford and the brief, five-page paper that followed would prove to be among the most influential works in twentieth-century astrophysics. The physicist Kip Thorne calls it "one of the most prescient documents in the history of physics and astronomy."[20]

Many years later, looking back on his career, Zwicky himself called it "in all modesty . . . one of the most concise triple predictions ever made in science. I think even David Hilbert"—the influential mathematician whom Albert Einstein consulted while formulating his theories of relativity—"would have been pleased since, in his will . . . he had left us with the admonition to be brief in all writings and to try to present our life's work in ten minutes."[21]

Zwicky and Baade's theories were essentially the birth of high-energy astrophysics. They were relevant to everything from pulsars and quasars to supermassive black holes—all the awesome processes in the universe that made life here on Earth seem even more trivial and inconsequential than before.

Zwicky's bold ideas shocked the more conservative members of his fraternity, who were still trying to figure out the basic building blocks of atoms. Here was this upstart, who suggested that temporary stars are not just interesting nighttime entertainment and fodder for the fantasies of court astrologers. Instead, Zwicky asserted, they play an important role in phenomena on Earth.

It was many years before it became clear just how important supernovae were to life on Earth. Explosions in space billions of years ago created iron, found everywhere from the ore in the Iron Range in northern Minnesota to our bloodstream; carbon, the basis of all organic life, and oxygen, which sustains life on Earth. These discoveries were years off, but that didn't prevent the two theorists from making bold claims—many of which would hold up over the coming decades.

An article in *Time,* "Star Suicide," explained Zwicky and Baade's new theory. "If they are right, the old concept of the end of the world," of life "freezing to death under a cooling sun—must give way to the prospect of [being] scorched to death by a sun having a final fling before joining the stellar ghosts in the cosmic graveyard."[22]

The *Los Angeles Times,* on December 8, 1933, called the prediction, a week before the APS meeting, "probably the most daring theory on the origin of cosmic radiation."

Knowledgeable people waited for the verdict of one man in particular, the Nobel Prize winner Robert Millikan. The son of an Illinois minister whose family came to the New World before 1750, Millikan was elegantly midwestern in his manners and dress, preferring bow ties and parting his close-cut hair with the same precision as he managed his famous oil drop experiment to measure the charge of an electron. He had also been interested in cosmic radiation. In fact, he coined the term *cosmic ray.* After lugging his equipment to the top of Mount Whitney to see if there was more or less radiation up there, he at first concluded that the radiation came from a source in or on Earth. "The whole of the penetrating radiation is of local origin," he said.

The finding was controversial from the start, but the eager scientific press decided that Millikan had done it again. He had tamed the electron, and now he'd tracked the mysterious cosmic radiation to its lair. Some began referring to the radiation as Millikan rays.

This naturally upset an Austrian physicist named Victor Hess, who in 1912 had taken an electroscope up in a hot-air balloon and discovered that the radiation levels were not lesser, as they should have been if the radiation came from Earth, but greater at altitude (Earth's atmosphere protects us from the full force of the rays; if it didn't, no life on the surface of Earth would be possible).[23] This, Hess believed, proved that the source of the Millikan rays was not Earth. And that Millikan had no more claim to naming them than he had to naming a planet.

In 1936, the Nobel Committee settled the question once and for all by awarding Hess the Nobel Prize in Physics. Even though the Millikan rays disappeared, Millikan himself remained the foremost expert in America on the subject of cosmic radiation. As determined a scientist as he was an administrator at Caltech, Millikan gathered up his equipment and trekked to Lake Arrowhead in California and Lake Titicaca in Peru. With better equipment and more careful measurements, he discovered that Hess was right. Cosmic rays, he said, must be a product of "God's laboratories in the stars."

But where were those laboratories? Finally, at a joint conference of the American Physical Society and the American Association for the Advancement of Science in June 1934, Millikan announced his latest results, obtained from an airplane that reached the stratosphere. His new measurements, he said, were compatible with Zwicky's, and Baade's, assertions.

It was enormous vindication, although the question would continue to be debated long after both Baade and Zwicky had departed the scene. Only in recent years has it finally been considered settled that supernovae do in fact produce highly accelerated cosmic rays.

It also took decades to confirm the existence of neutron stars. Although the mathematics of it were shown to be correct earlier, it was the discovery of pulsars in 1967 that proved they existed. Pulsars are neutron stars that spin faster than the blades in a kitchen blender.

Even without final proof, Zwicky's ideas were an immediate hit with the public. At the time, when a transcontinental commercial flight aboard the new Boeing 247 consumed twenty hours, readers of newspapers and magazines were fascinated by the new scientific theories emerging around the world. The problem was that most people couldn't understand them. Einstein's relativity? Niels Bohr's quantum mechanics? Interesting, but could you give that to me one more time? The idea of a supernova, an exploding star putting on a deadly galactic show, on the other hand, captured the fancy and anxieties of the public precisely because anyone *could* understand it.

Zwicky became the darling of reporters everywhere. Quotable, accessible, if occasionally truculent when some writer called him a Bulgarian (he was born there but considered Switzerland his true homeland), he became the hot new cultural commodity.

Though the idea of exploding stars may have been floating around, it was Zwicky who made the concept, and the name *supernova,* as familiar as relativity. He always had a way with the tart phrase, as well as the boldness necessary to force it into public use, even on the rare occasion when these attempts failed. His term for black holes—"Objects Hades"—was a much more colorful term that manages to convey both the uniqueness of the objects and the hellish conditions that prevail inside them. Despite his repeated usage, however, that name never caught on.

Popular magazines noted his birthday along with those of movie stars. He even made it into the funny papers. In fact, he said that he really only understood his new country when he was portrayed as a character known as Doc Dabble, a small, mustachioed man who looked a bit like Dagwood Bumstead's boss, Mr. Dithers, in the popular comic strip *Blondie.* In the January 1934 panel, Doc Dabble was shown bedding down in his observatory while his assistant sat at the telescope. "Call me when something explodes," Doc says.[24]

"And now I understand Americans," Zwicky said. "I am in the funny papers."

A national magazine chose him as the only representative of science featured in a column entitled, "They Stand Out from the Crowd." Selected with him were a poet, a playwright, a Soviet leader, and Mary Van Kleeck, a social reformer who fought to improve working conditions for women.

Those who knew Zwicky must have been amused to read the description of him as "modest and retiring when it comes to publicity." On the other hand, the writer said, he "has no hesitation in tackling the problem of explaining how it comes that in some types of stars a cubic inch of matter weighs a ton." His theory, *Literary Digest* said, is "now engaging the attention of a large part of the scientific world."[25]

True enough. Supernova talk was everywhere. Zwicky's transformation from modest college professor to scientific savant was fully under way. His life, and the universe, would never be the same.

On Christmas morning, 1933, newspaper readers around the country awoke to the suggestion that the Star of Bethlehem that supposedly heralded the birth of Christ was likely a supernova. Quoting a hymn that said the star could be seen both day and night, Zwicky told a reporter from the Associated Press that only a supernova was bright enough to shine both day and night.

"Still another coincidence," the AP report continued, was the biblical account of the ten-day journey by the three wise men. The star was said to have disappeared shortly after their arrival in Bethlehem. "Ten days is the normal life of a super-nova," the story said, incorrectly.[26] It's actually more like a month. No matter. Thus was born a story that would be told over and over in succeeding decades—becoming what is known as an "evergreen" in newspaper parlance—of how the Star of Bethlehem was probably a supernova.

The opportunity to further spread his ideas presented itself when radio came calling. The decade of the thirties has been called the golden age of radio, and for good reason. The number of American homes owning a radio more than doubled over that decade, from 12 million to 29 million. Foreshadowing the conquest of the American household by television, the squawking box became the central piece of furniture in every living room, the place where the family gathered to follow the adventures of the Lone Ranger and the Whistler. Zwicky titled his talk "Stellar Guests," the name given to "temporary stars" by the ancient Chinese. Appearing on WCAO in Baltimore on Tuesday, January 8, 1935, at 4:30 PM, just before "Jack Armstrong" and "Buck Rogers in the 25th Century," Zwicky's address was among those early scientific seminars on radio that allowed the average listener to hear great thinkers discuss their ideas in their own words.

Any star could be a supernova, Zwicky said, so the end of the world could come with the sudden explosion of our sun. When that happens, "this friendly planet would be nothing but a cloud of hot gases drifting in space."

To soften the blow, he told the story of a woman who asked an astronomer, "How long did you say the sun would go on shining?"

"Oh, a few billion years," he said.

"Thank heavens," the lady responded, with relief. "I thought it was only a million years."[27]

The lecture got a lot of attention. Accounts of it appeared as far away as Australia and London. Zwicky sent his Swiss friend, Rösli Streiff, a printed version.

She wrote back in stunned surprise at his emergence as an expert on the stars. She expected that since he was working at Mount Wilson, he would be up to date on the subject, "but it is new to me that you have a name there . . . I wouldn't be surprised to soon discover that you had become a surgeon or an outstanding statesman! Who knows, perhaps a medical examination would reveal that you have two brains, and in your land of unlimited possibilities, that would no longer be abnormal."[28]

✳ 2 ✳

GLARUS THRUST

LIKE ALMOST EVERYONE who discovers something great, Fritz Zwicky was fortunate to be alive in a transformative time. As late as 1922—before the cosmos became vast, before it became a cosmos—some learned people still thought our little neighborhood in space was everything. Distant galaxies and giant clusters of galaxies, neutron stars and black holes, pulsars and other strange creatures of deep space were yet to be discovered.

The men and women who found these things were like explorers in every age: adventurous, determined, egoistic. They did not sit on safari elephants but at the controls of giant telescopes, riding the crest of light waves into the birth chambers of creation. The creatures they brought back astonished the world, turning them into celebrities and superstars. When Einstein landed in San Diego, he was mobbed by three thousand schoolchildren bearing flowers. People didn't understand relativity, but they wanted to believe in a future when science would solve all the nasty problems that had harassed humankind for ages. Zwicky and the other famous savants, as they were called then, were consulted on everything from whether there was life on Mars to the existence of God.

Among the explorers of the universe's darkest continents, Fritz Zwicky was one of the boldest, if not *the* boldest. He cared little for the opinions of experts. Some of his ideas sounded crazy, like pelting the moon to see if there was water, something NASA finally did in 2009. Some of his ideas—such as the Terrajet, a Lex Luthoresque machine designed to chew through the Earth to erupt in the capital of an enemy—probably were crazy.

Even in his lifetime, Zwicky was something of a throwback. The time of the enigmatic genius laboring alone in his or her laboratory was coming to an end. The lone wolf—Zwicky's favorite term for himself—was being replaced by staff committees and boards, each obliged to a different institution. Every individual would be responsible only for his piece of the whole, whether it was unraveling DNA or studying rocks on Mars. Today, the typical NASA planetary mission employs hundreds of scientists. One is expert in riparian environments, the next knowledgeable about aeolian forces in deserts. Over there is the world's leader in the preservation of biomarkers, and so on. Fritz Zwicky saw all this coming and hated it.

When he was old and famous enough that people began asking him about his influences, Zwicky recalled his father giving him two pieces of advice when he was a boy that he carried with him always. If he should pursue a career outside Switzerland, he should be ready to sacrifice a substantial portion of his earnings to the government without complaint. This was a citizen's obligation in a civilized nation, Fridolin Zwicky said, and must not be shirked. Second, he told his son, "life would not be worthwhile if I looked left and right for what others thought about my ideas and actions or gave up any ounce of my independence."[1]

It was the sort of thing any father might say to his child. *Be your own person. Stand up for yourself.* Young Fritz took to the words as if they had been handed down on stone tablets. When embroiled in one or another of his pitched battles with colleagues or government ministers, he would declare that their demands were not something any "free and independent" person could possibly tolerate. Free and independent—alternatively, "lone wolf"—was how he typically characterized himself when he was stubbornly refusing to fall in line. And it didn't matter whether the issue was the assignment of an office at Caltech, the editing of a paper outlining a new cosmological theory, or the price of a meal in a restaurant.

Fritz Zwicky would always be ambivalent about his homeland. When he became famous, a man whose opinions were sought on subjects far beyond his expertise as a scientist, he wrote glowingly of how the Swiss had had a standing army for centuries yet avoided the militarism that entangled the great powers in foreign adventures. However, he disliked the insularity and smugness of the Swiss, even, with the kind of grouchy hyperbole he became known for in old age, calling them "the stupidest people on Earth." It was an insult his fellow Swiss never forgot.

Zwicky's ambivalence is fitting, because Switzerland is ambivalent about itself. It's never quite sure that it even wants to be a country. In America, people remark on the power states have to ignore orders from Washington. This attitude is doubly powerful in Switzerland, where twenty-six cantons each have their own constitutions, welfare policies, and legislatures. Until 1848, when the Swiss Confederation was formalized, each was a sovereign state. Of those states, Glarus, the ancestral home of the Zwicky family, is among the oldest, dating to 1352.

It's also among the smallest, at 264 square miles. Yet it holds an outsized place in Swiss history. One of the original thirteen cantons, it was once a forest region of herders and builders. The Alps dominate Switzerland, but they are particularly imposing in Glarus, a narrow valley carved out of the limestone by ancient glaciers. Within twenty-five miles, the valley descends from the 12,000-foot Tödi, a mountain massif in the south, to the swampy heel at Bilten, 1,400 feet above sea level.

The briskly flowing Linth River splits the agricultural valley. On either side, the Alps rise sharply, making up fully a third of the canton. These mountains contain one of the most important geological structures in the world. Known as the Glarus Thrust, the strange conglomeration of rock mystified observers when it was first described in the nineteenth century. It was clear that older, more degraded strata lay on top of younger sediments. How could this be, early geologists wondered. It was as mystifying to them as it would be for archeologists to find the remains of a Roman villa lying on top of a Dickensian workhouse. It was only when scientists discovered plate tectonics that the mystery was solved. Continental plates colliding deep in the Earth force old rocks up and over a younger surface: This is the key to mountain-building the world over. Today, the Glarus Thrust is a UNESCO World Heritage site; a model of it is housed in the American Museum of Natural History.

The beauty of its mountains and river belies Glarnerland's bloody past. In some portraits, St. Fridolin, the itinerant Irish monk who converted the people of northern Switzerland to Christianity in the sixth century, is shown holding a staff, a simple man of God bringing civilization to heathen tribes. In others, he is shown walking arm-in-arm with a skeleton. The story is that as a rich landowner named Ursus lay dying, he tried to curry favor with the Christian God by donating his land to Fridolin. Ursus's brother, Landolf—not so near death and less in need of the monk's intercession—sued to invalidate the bequest. Untroubled, the priest exhumed Ursus's decayed body to stand witness in court. Landolf was so horrified by the monk's audacity that he turned over his part of the inheritance as well.

The Habsburgs of Austria ruled the region until the late fourteenth century, when the Glarners rebelled, leading to a battle that is still celebrated, more than six hundred years later. A Habsburg army of 7,000 foot soldiers and horsemen attacked the town of Näfels just down the road from the canton's capital, also called Glarus. Breaching the defensive wall, the army stormed into the valley, looting and burning with true medieval enthusiasm. As the story goes, the men of Glarus fled into the snowy mountains and huddled around St. Fridolin's banner. The invaders discovered them and attacked, only to meet a hail of stones, boulders, and even tree trunks hurled from above. The riders were unhorsed and the soldiers dismayed, retreating in panic with the small army of Glarners in pursuit. As the Habsburg force fled over a bridge on the Walensee, the structure gave way. The proud invaders plunged to their deaths, defeated by a few hundred villagers. The fifty-five Glarners who lost their lives in the fight were buried with pride in the cemetery at Mollis, the same revered plot of land that would one day hold the ashes of Fritz Zwicky.[2]

Until about 1800, the local economy was anchored by the money brought home by Glarus's mercenaries, who earned a reputation for courage in foreign wars. "In the past ages the Swiss were regarded as the best, bravest, bloodiest and most reliable fighters in existence," wrote Max Eastman.[3] So great was their reputation that Glarners originally formed the Pope's Swiss Guard. After mercenary service was outlawed, Glarners became tillers and traders, exchanging local slate and wool for exotic goods from the east. Glarners like the Zwickys earned a reputation for astuteness and determination. A well-worn saying holds that in every trading center, "one meets a Jew and a Glarner."[4]

The town of Glarus, where Fritz Zwicky was raised, is one of the canton's largest cities, though "large" is a deceptive term when the population of the entire region is less than 40,000. The town's landmarks include the Romanesque Burg chapel on a small hill overlooking the city, which dates back a thousand years. The most imposing structure is the twin-towered Protestant church that dates to shortly after the fire that destroyed much of Glarus in 1860. Its triple-naved sanctuary rises majestically above a dwindling number of worshippers. The Swiss these days are not particularly devout, although they continue to honor conservative values. Shops are shut tight on Sundays, and signs posted in shop windows urge children to respect their elders.

The avenues of Glarus are narrow, traffic can be a hazard to the unwary, and there are few first-rate restaurants to attract the international tourist. But the ever-present sense of history, combined with the unmatched scenery of the great Glärnisch mountain, looming over the city the way a big, protective dog watches over an infant, can be compelling at any time. Whatever the frustrations of life, all one has to do is look up, and the mountain's spectacular presence never fails to produce a catch in the throat. This is the place that nurtured a young exile from Bulgaria and fed an already substantial sense of pride in place and family.

Born on February 14, 1898, Fritz Zwicky was the eldest of three children of Fridolin Zwicky and his Czech wife, Franziska Wrcek. The Zwickys were an old family, going back generations in the mountainous region of northeastern Switzerland. As a young man, Fridolin moved to Varna, Bulgaria, in search of business opportunities. Varna before World War I was a cultural crossroads with thriving Jewish, Greek, and Roma communities. Its setting on the Black Sea made it a favorite vacation spot for those escaping northern climates. Fridolin set up business in this open-minded resort, where the high temperature in late July rarely exceeded 82 degrees, and tourists could choose from twelve miles of fine sand beaches. Besides managing the Swiss Credit Bank in Varna, Fridolin Zwicky sold Swiss textiles and machinery, later opening a lead shot factory that supplied ammunition to hunters.

The elder Zwicky was so scrupulous in his dealings that it was said of him that the "day that the Zwicky company does not pay up will be the end of the world."[5] A half-dozen different governments, including Norway, Germany, Austria, and Turkey, chose him to represent their consular offices in Bulgaria.

Fridolin Zwicky's rigor and rectitude cast a long shadow over his son's life. His example was almost certainly the root of Fritz's rugged individualism and lifelong hatred of Communism. When the Communists seized control in Bulgaria after World War II, the Zwicky family would lose the business and the family home. Fritz Zwicky, already skeptical of Russian socialism, became a vocal campaigner against Communism in America, railing against scientific colleagues sympathetic to the red cause.

Even so, Fritz believed that the core of his being, his imaginative genius, was a gift from his mother. At a time when women in the Balkans were expected to take a subservient role, Franziska Wrcek was a force of nature. Her name was well known in the Varna bazaars; she knew the languages of all the traders and was unafraid to go nose-to-jaw with the merchants in pursuit of the best price at a time when "it was only men who haggled," Fritz wrote later.[6] She staked out the territory early. On the day after her wedding, she returned to the bazaar, leaving her new husband home to ponder his future married life.

Her son took her example so much to heart that when he was a professor at Caltech, his nightly bedtime reading was *Quinta Linguas,* a magazine that translated every article into five different languages. One of his proudest accomplishments would be making the first broadcast in Russian, which he taught himself, over the Voice of America. The subject was the achievements of American science.

It might seem strange that he absorbed these lessons so thoroughly, since he spent so little of his youth in Varna. When he was six, his father sent him away to school in Glarus, where he lived with his grandparents in a tidy, three-story row house on Burgstrasse, a stone's throw from the railway station. Fritz's brother Rudolf, two years younger, stayed in Varna, as did their younger sister, Leonie. Fridolin Zwicky's plan was for his oldest son to get a sturdy Swiss education in business affairs, then return to Varna and take over the family enterprises.

Rather than alienate him from his family, the experience consecrated his early memories of his father's probity and his mother's hardheadedness. That he missed his family and tended to idealize the life he lost in Varna was clear from his writings. As an adult, Zwicky was not much given to self-analysis or memoir writing. But a train trip home to visit his parents in 1910, when he was twelve, inspired a rare poetic streak.

Portrait of the Fridolin Zwicky family, taken in Varna, Bulgaria, in 1915. Fridolin, businessman and diplomat, is at left. Standing at the rear is Fritz, the first-born, a teenager here, and his younger brother, Rudolf. At right is their mother, Franziska. Their sister, Leonie, is seated in front.

He was aboard the Orient Express, the famed rail line associated with luxury and international intrigue. Twilight was falling, and he was feeling a sense of peace that would be rare in his life. "It was early summer and the remaining snow on the peaks of the Tirol glowed pink in the rays of the setting sun, before the dark shadows of the valley reached up and captured them. I was twelve and it was holiday time."[7] As he left behind the lakes and forests of Switzerland, he mused, "The people were friendly and I loved the mountains, so this made up for the absence of my family, at least a little."

The train rolled through Vienna, Budapest, Belgrade, and Sophia, the ancient, exotic cities of eastern Europe arrayed along the 1,770-mile-long Danube, on the banks of which Alexander the Great rested his army. The boy fell asleep on his luggage and dreamt of a "carefree future. The whole world lay before me in all its beauty, with all its amazing potential and revelations," he wrote.

This golden future lasted only two years, before the outbreak of the Balkan Wars of 1912–1913 ushered in the bloodlust of the twentieth century, the deadliest period in all human history. In the First Balkan War, Montenegro, Greece, Bulgaria, and Serbia teamed up to scatter the decayed remains of the Ottoman Empire. In the second, Serbia, Montenegro, Greece, Romania, and the Ottomans of Turkey ganged up on Bulgaria. In the two wars, 224,000 people lost their lives. It was just a prelude to the horrors of World War I, unleashed by a subsequent Balkan crisis.

"So far as my Swiss friends knew, these wars might have been fought on another planet," Zwicky wrote. For his family, however, the Balkan Wars were the first blows in a series of conflicts that would rarely leave them in peace for the rest of Fridolin's life. "The Balkan war meant the beginning of all the tragic events that were to leave me alone, with my Orient Express visions of the future shattered. I spent many long years trying to find a way of recapturing this lost world."

The restless discontent left by those stolen dreams was something that drove Fritz Zwicky as a scientist. It goes a long way toward explaining his refusal to confine himself to one discipline, even when that discipline, astrophysics, made him famous.

In comparison to the travail in the rest of Eastern Europe, Zwicky's life with his grandparents was an island of calm. But it was an isolated existence. His grandparents were kindly but too old to be companions. And the house, which had been built as workers' lodgings during Glarus's textile boom years in the nineteenth century, was too small to confine an active boy.

It was worth remembering, he wrote in *Jeder ein Genie* (Everyone a Genius), "that I lived alone in Glarus with my old grandmother and because I did not have a brother or sister to play with, I would go off playing in the mountains or play cards with policemen and foresters."[8]

The card game was jass, which is something of a national obsession in Switzerland. Dating to the eighteenth century, jass is similar to—and is said to have influenced—the American card game of pinochle. Instead of hearts, clubs, diamonds, and spades, the suits in jass are roses, acorns, shields, and bells, all artfully rendered in a style reminiscent of medieval iconography. Despite, or perhaps because of, its old-world feel, it's so popular today that it's not unusual to see couples finish off an expensive restaurant meal with a game of jass over glasses of crisp Swiss wine.

The boy's analytical mind was completely absorbed by the game. One of his favorite places to play was at the police station on Burgstrasse, just down the narrow road from his grandparents, under the looming Glärnisch mountain. He soon bested the constables and went on to become a local champion at the age of nine.

His father sent his oldest boy away to learn business, but Fritz's teachers at the Höhere Stadtschule (grammar school) in Glarus noticed that he had an unusual aptitude for mathematics. They wrote Fridolin that it would be a shame if the boy's talents were lost to commerce. "My father wrote me that in his opinion my brother was the real crook . . . and I should go into engineering or as a scientist,"[9] Zwicky told an interviewer many years later.

A childhood friend, Jakob Stähli, recalled Fritz's aptitude and energy at school. "He excelled in every subject" without trying.[10] An early essay arguing for the construction of a local swimming pool so impressed his teacher, the rector Nabholz, that he arranged to have it printed in the town newspaper. The pool was built, but only much later.

Fritz's intelligence impressed his teachers, but his vitality was what drew friends and playmates. He was the only boy that Hans Hefti, another childhood companion, recalled clearly from his youth. "He was solid, well built and strong," Hefti said later. Four feet of snow falls in Glarus each winter, which provided Fritz and his friends with diversion and the opportunity to display their physical prowess. During snowball fights, everyone knew the safest place to be was at the side of Fritz, the biggest boy and by far the most reckless and headlong in his enthusiasm. "Stick with Zwicky" was the motto adopted by the other boys.

By the time he reached secondary school, Fritz had grown into a strong, confident boy. In adulthood, and at the peak of his powers of penetration, Zwicky had a habit of standing with his feet apart, his large head cocked in a way that seemed to say he was carefully assessing all that he was observing and would render a final judgment at the appropriate time. This confident assertiveness was already established in the teenaged Fritz. Unlike in later years, when he sometimes seemed to go out of his way to make mischief, he was well liked, with a magnetic personality. "He was not a loner or an eccentric but very good to be with," Stähli told Zwicky's Swiss biographer.

In those years, Fritz gathered around him a group of admirers who met on sunny afternoons in Volksgarten Park, located just across the road from the gingerbread architecture of the railway station. In winter, the station is crowded with skiers headed for Glarus's slopes; the park today, as it did then, attracts strollers, lovers, and park bench philosophers. Though he was friendly and could be a fine storyteller in adulthood, it was noted then, as it would be later, that Zwicky had no facility for nor interest in small talk. For him, words were instruments, tools, useful to achieve an end, not simply to smooth social interaction and build friendships. Alfons Kubli, a year behind Fritz in school, was too intimidated to participate in the lively debates Zwicky organized. "I never aspired to the philosophy bench," Kubli said dryly, later on. "But I can still see it clearly: Zwicky with his distinctive head, like a modern Socrates surrounded by his disciples."[11] In these debates, and in his writings at school, he drew on his wide reading, from Tolstoy to Faraday, from psychology to cowboy stories.

In 1914, with the world on the precipice of world war, Fritz was on the move again. This time he went to Zurich, the bustling capital and crossroads of both ideas and nationalities, to attend the Industrieschule, the Zurich technical college, reputed to be the oldest and strictest school of its kind in Switzerland. There he met a boy who was destined to become a lifelong friend, climbing partner, and fellow scientific adventurer. Young Tadeusz Reichstein was every bit as gifted as Zwicky, if perhaps not as iconoclastic in his interests and explorations.

Reichstein was from Poland, the studious, bespectacled son of an engineer and a few months older than Zwicky. History remembers him principally for developing the process of synthesizing vitamin C, one of the body's most important nutrients. The Reichstein Process allowed the vitamin to be produced on an industrial scale. In 1950, he won the Nobel Prize (in Medicine), an honor his robust Swiss friend would never receive.

Reichstein was one of the few people that Zwicky regarded as both an equal and a confidant. The letters they wrote over the years bore a familiarity and a modesty that were absent from most of Zwicky's correspondence. In 1952, when Zwicky was a world-renowned scientist, he wrote to his friend in simple, honest frustration: "In astronomy I was fairly lucky in discovering vast amounts of intergalactic matter," not bothering to mention his far more important discoveries of dark matter and neutron stars, both of which had yet to be proven correct.[12] "Too bad there seems to be no quiet empty place where one could retire and get away from the horse faces."[13]

Later on, Reichstein recalled his first encounter with Zwicky. "I had always been good at maths," he told an interviewer. "When the examiner came, he asked a few questions and Zweifel (another student from Glarus) answered, producing succinct answers, as did Zwicky, all correct. I asked myself, 'How could they know all that?' It was only afterwards that I noticed that actually they were both really able at maths."[14]

Along with playing cards, shorthand is something of a national passion in Switzerland, both the doing of it and finding ways to improve on it. They call it "debate writing," of which there are two types. The shorter, more difficult one is called *debattenschrift*, which Zwicky quickly mastered. Then he went on to invent his own, even shorter and more difficult form of debate writing.

"Being large boned I could not write as fast as some of the others," he explained, "so I embarked on my first morphological study."[15] Morphology is a system of thought that Zwicky invented later on. The basic idea was to approach every problem with a completely open mind, analyzing all possible solutions, no matter how far-fetched. He claimed the method helped him arrive at some of his greatest insights, ideas that helped turn a little Pasadena rocket company, Aerojet, into a billion-dollar Cold War enterprise. It also lay behind his greatest follies, such as proposing to move planets around the solar system to make them habitable for human beings. Forgotten now by all except a few devotees, the same kind of people who still believe Esperanto is the language of the future, morphology was once considered so promising that the US government invited Zwicky to submit a proposal to teach it to military officers.

"Why only shorten characters?" Zwicky wondered. "Why not abbreviate every word, say a few thousand instead of just the few dozen in the standard shorthand system? And why not use symbols to represent sentences which keep occurring and even sentences of sentences? In this way I reduced the sheer bulk of shorthand script so at the age of seventeen I was the Swiss champion in both French and German." He called his technique double double shorthand, "with which you can write faster than you can talk."[16]

In school, he and Reichstein worked out an arrangement. Zwicky would record the lecture in his invented shorthand. When the instructor used a word that was new to him, or outside his system, he elbowed Reichstein, who wrote it down in longhand. In this way, the two could keep up with even the most energetic lecturers.

Clever as he was, there was a problem with this technique. Because he never taught the method to anyone else, many of Zwicky's diaries that use the system are, to the dismay of students of scientific history, virtually unreadable. Dozens of them lie in boxes and folders in Zwicky's archive, which takes up almost half of the chilly basement of Glarus's main library. Elsewhere in the sprawling archive sits a piece of scorched wood from a temple near Ground Zero in Hiroshima, retrieved during a secret trip to Japan for the military after the war, along with his notes from a similar trip to Germany, where he probed Hitler's top-secret rocket works at Peenemünde and Völkenrode.

Reichstein was Zwicky's first roughnecking friend. No one who spends any time in Glarus can fail to appreciate the way the mountains impose themselves on daily life. They stand over everything, defining and explaining the landscape, inviting the contemplative to wonder and the adventurous to climb. Fritz Zwicky caught the mountaineering bug early.

He appreciated the beauty of his childhood surroundings, but for him, conquering them was where the true reward lay. A lifelong member of the Swiss Alpine Club, he wrote often about his trips aloft. Over six feet tall and possessing a strong constitution, he regarded mountaineering as the true test of a person's worth.

The Alps, though not the tallest mountains in the world, contain some of the most challenging ascents in Europe. One has only to study the rugged face of the Glärnisch to appreciate the difficulties, especially for those like Reichstein and Zwicky, who were never satisfied with simply making the climb. They wanted to find, and master, the most difficult way up.

They had plenty of opportunity for trail blazing since mountain climbing for recreation was a relatively new pursuit in Europe when Zwicky and Reichstein were young. Its beginnings are often traced to the first summiting of the Matterhorn in 1854. The Swiss Alpine Club had only been founded in 1863. Even in America, when Zwicky announced

The north face of the 9,651-foot Glarnisch mountain, where the young Fritz Zwicky honed his mountaineering skills. Why do climbers keep "madly racing up mountains?" Zwicky asked. His telling answer: It is "something no one can take from us."

his plan to climb Mount Whitney in 1927, there were no records that anyone had previously attempted a winter climb.

Reichstein, who died in 1996, recalled their treks together. "We usually stayed in the Glarnerland region,"[17] he said. "It was still affordable then. You could get by with five francs [now about $6] for a weekend . . . the hut charge was fifty centimes. We took our own firewood up with us." The only problem was the rain. All-weather clothing was unknown, so the two were often soaked to the bone by the end of the weekend. But both were undeniably intrepid.

While still young men, Reichstein and Zwicky climbed the 10,719-foot-tall Clariden, the challenging Bocktschingel, and other mountains in the Glarus Alps, which Zwicky came to know intimately. But the 9,651-foot-tall Glärnisch was his favorite. The Glärnisch has several peaks, the major part of which consists of two ridges separated by the Glärnischfirn Glacier. The glacier has lost much of its mass in recent decades, but in Zwicky's youth, it was a mile long, a half-mile wide, and tricky to cross, with crevasses suddenly appearing at one's feet. From the peak, the climber can see the Glarus valley and Lake Klöntal, a spot of such remote beauty that, in the words of Nobel Prize–winning author Carl Spitteler, it is "unimaginable even in dreams."

Climbing the Glärnisch was not too difficult if one used the regular route, but Zwicky wanted to try an ascent up the supposedly impossible north face. "We were always talking about [it] and would say, 'One day we'll do the north face,'" Reichstein recalled.[18] The route is exposed and steep enough to induce vertigo. After laborious climbing over the rock face, one encounters scree (loose rocks) almost to the peak. In 1924, the friends finally became the first to reach the summit by that route, a climb Fritz recalled often in subsequent years. They didn't stop there. Barely two weeks later, the pair made the first ascent of a prominent tower on the west ridge of the mountain, christening it—as was their right—with the name it still bears: Glärnischnadel—the Glärnisch Needle.

ON THE TRAIL OF ZWICKY'S GHOST

The Bones of the Earth

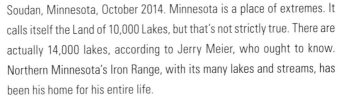

Soudan, Minnesota, October 2014. Minnesota is a place of extremes. It calls itself the Land of 10,000 Lakes, but that's not strictly true. There are actually 14,000 lakes, according to Jerry Meier, who ought to know. Northern Minnesota's Iron Range, with its many lakes and streams, has been his home for his entire life.

It's hard country, and Meier, a big man with a spread around the middle that might have come from eating too much fried walleye and hunkering down for months during the brutal winters, has weathered, and been weathered by, its extremes. In the small town of Tower, the thermometer registered a temperature of −60 degrees a few years ago. Until then, it was the coldest temperature ever recorded in America east of the Great Plains.

It was even colder in Embarrass, where Meier lives on three white pine acres. The town's official thermometer shattered in the arctic chill that night, so no one knows how much colder it got. "What was it like?" he replied, shrugging. "Pretty much like any other night when it's thirty below or colder. You stay in the bloody house. The thing I couldn't believe was the idiots from the city who came up here in snowmobiles and built igloos to spend the night."

When he referred to the city, he meant Duluth, or Minneapolis, 243 miles to the south. Up here on the Iron Range, a challenging canoe paddle from the Canadian border, they are amused by the strange doings of people who live in cities. They enjoy playing practical jokes on them, such as the tongue-in-cheek campaign the little town of Ely mounted to host the Olympics. City fathers announced they'd already drained Miners Lake for stadium seating. Outsiders bought the ruse, or pretended to. Tee shirts adorned with the slogan "Ely in 2016" became popular around the state. Billboards with the slogan showed up on the interstate.

But about some things the people of northern Minnesota are deadly serious. Meier grew up in the area. His father worked construction; his mother was a nurse. Both of them grew up here, too. All those hard winters still don't make him a native in the eyes of old-timers. You've got to be able to point out your great-great-grandfather's grave before a "pack sacker," as outsiders are known, becomes a true Iron Ranger. That's the title of honor conferred on the polyglot immigrant population that streamed in from Slovenia, Finland, Norway, and other frozen lands to work the mines a century ago.

Meier supervised one of those old pits. The Soudan Mine, the deepest ever dug in Minnesota, at 2,361 feet, once produced some of the purest iron ore in the world. From the late nineteenth to the mid-twentieth century, they pulled millions of tons out of the ground—two million in 1902 alone, when U.S. Steel bought the mine—and shipped it off to the mills in Pittsburgh. There it was turned into fine grade steel for the skyscrapers and sleek automobiles that heralded America's emergence as an industrial colossus. Old Iron Rangers contend that modern steel, made from lower grade ore, can't touch what came out of Soudan. In its heyday, Soudan, which, in another fit of local whimsy, earned its name by being as inhospitable in winter as the other Sudan is in summer, was known as the "Cadillac of mines."

There once were 1,800 mines in this part of the Iron Range. But as it will, progress came. The mining companies figured out how to use magnets to separate good iron ore from poor quality ore, known as taconite. That meant they could scoop up everything, 300 tons at a time, rather than struggle to dig out only the best ore. That was the death knell for Soudan. On December 15, 1962, the hoistman hauled up the last pure iron ore that would ever come out of the Cadillac of Mines.

In recent years, another form of mining had been under way at Soudan. It was the skeleton of the universe they were seeking rather than the bones of the Earth. The folks at the bottom of the mine were chasing Fritz Zwicky's ghostly particle, dark matter.

✲ 3 ✲

THE BIGGER AND BETTER ELEPHANT

BY THE SUMMER of 1916, the Great War—the first one, the "war to end all wars"—had been raging for two years. Fritz Zwicky was eighteen years old, tall, handsome, with a round, moonlike face and dark, penetrating eyes that gave him an almost vulpine expression of curiosity. Despite the devastation in the wider world, the young Swiss had every reason to feel good about his future. He had mastered hard mountains, and his final report at *Industrieschule,* covering fourteen subjects, awarded him a score of 82.5 out of a possible 84 points. He may have been fudging when he said that no one in the past 150 years had achieved such a score. Even discounting bluster, it was a notable result. More revealing was the fact that he had done it without bothering to study. Learning, like so much else, came so easily to him that he took it for granted.

He saw, without quite knowing why, that he approached problems "in a way that was rather different from my friends." That his early teachers rewarded him was a credit to their open-mindedness, a characteristic he would come to feel was missing in America. "It was exactly the opposite with my scientific colleagues, and it was decades before at least a few of them came to agree with me that there was more to the morphological approach than might appear at first sight."[1]

Zwicky's early success convinced him that there was a purpose awaiting him. As yet, he didn't know what that purpose was, but the bloody violence at Verdun, the Somme, and Ypres already convinced him that his particular gifts were needed in a world in flames.

In October, he enrolled with his friend Tadeusz Reichstein at the Swiss Federal Institute of Technology. One of the elite universities in Europe, ETH is still considered one of the world's leading universities in the fields of engineering, science, and technology. Graduates hold twenty-one Nobel prizes. Its most illustrious graduate is Albert Einstein, who was a generation older than Zwicky. Later on, American journalists often called Zwicky Einstein's "most promising" pupil; Zwicky used similar terminology himself. Einstein was a professor at ETH but had left for a professorship in Germany by the time Zwicky arrived. Likely the only way he could have taught Zwicky was by paying a visit to the Industrieschule to tutor Zwicky, which one early American newspaper account said is what happened.[2] Whether or not he ever taught Zwicky directly, Einstein cast

Zwicky with his childhood friend Tadeusz Reichstein, in Basel, Switzerland, in 1959. The Nobel Prize winner was one of the few people Zwicky regarded as an equal, both as a scientist and as an outdoorsman.

a long shadow. Throughout his life, the Swiss could not resist comparing himself and his accomplishments to those of the great man, even going so far as to paste his and Einstein's entries from "Who's Who" books side-by-side into one of the many pocket diaries he carried in his suit jacket. Into those pocket-sized notebooks went everything from equations for the mass of galaxies to errant philosophical observations and incitements to still greater action and accomplishment.

In Zurich, Zwicky found himself in one of the most dynamic and stimulating cities in the world, in that or any other century. In the first decades of the twentieth century, Zurich exerted an irresistible pull on every kind of émigré artist, crazy revolutionary, and vagrant castaway in Europe. They were dharma bums before the dharma left India and hit the road. The poet, Franz Werfel; the writer, William Somerset Maugham; and Erich Maria Remarque, author of one of history's greatest antiwar novels, *All Quiet on the Western Front,* all pitched up there. James Joyce, a habitué at the Café Odeon, where the Dada movement began, spent five years in Zurich, working on the gloriously impenetrable *Ulysses.*

Political agitators and exiles sat for hours in local bars and bratwurst places scribbling socialist position papers to be published in pamphlets and newspapers that almost nobody read. Benito Mussolini was there, along with Leon Trotsky and other troublemakers.

Surrounded by larger nations at war, the Swiss could not help but look over their shoulders for fear that they, too, might be drawn in.[3] One of the loudest and most pugnacious of the émigrés trying to make that happen was a small, fussy man in a bowler hat and oily greatcoat who had dreamt for years of founding a true workers' state. A place where the bourgeois imperialists—of which Fridolin Zwicky was undeniably one—would not just be toppled from their self-satisfied perches but hanged from lampposts should they object too loudly.

Vladimir Ilyich Ulyanov, known to the larger world as Lenin, lived with his wife and confidante, Nadya Krupskaya, in a small, shabby apartment on Spiegelgasse, not far from the university where Fritz Zwicky was studying mechanical and electrical engineering. In later years, Zwicky would often say that Lenin lived next door, but he likely meant in the same scruffy neighborhood of bohemian apartments that appealed to students surviving on limited funds. Zwicky would never have wanted to hear it, but he and Lenin had a lot in common. Like Zwicky's father, Lenin's rose to wealth in business. Both their mothers died when they were young men, and both were deeply mourned. Both men found little to admire in traditional religion and came to trust their own instincts so much above anyone else's that they often found themselves isolated. But where Zwicky would fight his battles in the halls of academia and at the controls of great telescopes, Lenin's stubborn belief in a society where the workers made all the rules transformed the world.

In Zurich, Lenin surrounded himself with a small group of young idealists who met in the same small cafes frequented by the student Zwicky. At 47, Lenin was much older than his ragtag band. With his bald head, untrimmed red beard, and arched eyebrows, he made an unimpressive appearance. Until he began laying out political strategy. Most of his opponents dismissed him as a flaming sectarian with no real following. But Lenin knew it was better to have a united minority than a flabby majority. When the time came for action, the minority would strike and the majority would follow. The wind always blows from the left, he said. History would prove him right.

By the fall of 1916, as Zwicky was getting down to work at the Institute, Lenin was despairing of the Russian proletariat ever rising up against Tsar Nicholas. He adopted a new strategy: making revolution first in Switzerland.[4] With its three major languages—French, German, and Italian—Switzerland was perfectly situated to spread revolt everywhere. Unlikely as it seemed, and was, Lenin believed Switzerland could be the vanguard of the revolutionary war to come.

During the previous winter, there were food shortages in Switzerland. The number of foreigners, and especially the radicalism of the Russian émigrés, began to concern the authorities.[5] At Carnival, the wearing of masks was forbidden. Swiss agents began taking closer note of the foreigners' activities. Fearing expulsion, Lenin tried to conceal his movements. With his waterproof bag slung over his shoulder, he looked more like a bedraggled vagrant in his worn bowler hat, his unkempt beard and moustache clinging to his lean face like tree moss, than the man who would soon take the reins of the Russian nation.

His influence among the young firebrands, however, was unmistakable. Zwicky followed the street demonstrations of Lenin's followers in the *Jungburschen,* the socialist youth group, with growing alarm. One demonstration brought two thousand of them into the streets, waving red flags, singing the *Internationale,* and chanting "Down with war." On August 1, hundreds fought with the Zurich police.

Despite Lenin's efforts, however, the Swiss could not be moved out of their comfortable, sedentary lives. The nation of herders and merchants, who mowed the slopes of the mountains four times each summer, no matter how steep the ground, simply refused to cooperate. "Such warriors! Such leftists!" Lenin wrote in February 1917. The Swiss were "almost hopeless."[6]

Only a month later, on March 13, 1917, the tumblers of fate finally clicked into position for Ulyanov. Tsar Nicholas abdicated the throne, and a provisional government was established. By November, Lenin stood at the pinnacle of the new revolutionary government in Moscow.

The hoped-for Swiss revolution never took place. Fritz Zwicky played a small role in its failure. In a lecture many years later, he described watching Lenin and his agitators "at close range. . . . It was obvious to me from the start that they played one of the foulest and most murderous games of all history," Zwicky said.[7] "Therefore, when 1918 came, and even some of the Swiss laborers got fooled by slogans from Moscow, we founded a temporary political organization in Switzerland to smash Communism in our country once and for all."[8]

He left school for a time to become a secretary of the organization, the Federation for the Reforms of the Transition Time, meaning the transition from wartime to peacetime. The federation brokered relations between workers and employers to improve conditions. It was eventually disbanded, but according to Zwicky, the members stayed in touch and continued to work together over the years "for the alleviation of the suffering of children and old folks in distress."[9]

This experience showed him the world outside the academy. His organization could hardly claim sole responsibility for Switzerland's rejection of Communism; Swiss stubbornness and the success of more moderate social reforms did that. But his political activity proved that the world is too wide and wild with opportunity to confine oneself "to one particular vocation or avocation," Zwicky said later.[10]

Eventually, this insight would mature into a new belief system, which held that "any individual could be satisfied only with a profession which somehow closely fitted his own and very specific characteristics. This led me to the idea that every person is really quite unique, incomparable and irreplaceable or, in other words, each person has a potential genius which, if not properly developed, will result in frustration and unhappiness. . . . The proper profession for myself obviously still had to be invented. . . . I called my new activity morphology."[11] The study of structure, the term was lifted from biology.

For Zwicky, the greatest difficulty was not finding his own genius. He found that when he arrived in California and started watching the skies from the top of a classroom building at Caltech. The greater problem would be finding a way to help others find their genius. "Although I have tried hard, I never achieved criteria for finding out about or developing a man's genius," he said in middle age. "I am sure that each one is a very particular person and has a very particular ability and particular visions, but I don't know yet how to recognize it."

Even if he couldn't lead the way for others, he saw all around him the destruction caused by those who never found their way. "Some men became so frustrated that they ended as mass murderers, such as Hitler, Lenin, Mussolini, Stalin, and so on," he said. "They went completely astray and were miserably unhappy because of being beset by inferiority complexes."[12]

"These dictators apparently never had one sunny day or even a single sunny hour in their lives."[13]

Lenin was indeed unhappy in Zurich, laboring to bring about a revolution no one wanted. But it seems unlikely he would have agreed that he failed to find his genius. His talent for political organizing and turning opportunity to victory saw its full expression in the 1917 Russian revolution.

But Zwicky's thoughts about the cause of human unhappiness held more than a nugget of truth. His insight anticipated the rise in the middle of the twentieth century of the therapy industry with its assemblage of counselors, gurus, and street-corner proselytizers. They would all wrestle with the problem of how to inspire others to find their purposes. It was clear to Zwicky, and it would become clearer as the century drew on, that people needed meaning in their lives as much as bread.

Though he began his student years in the engineering department, when he returned to school after his political work, he switched disciplines, studying for a teaching certificate in maths and physics. For his final examination in 1920, his maths examiner was Hermann Weyl, one of the greatest mathematicians of the century. The highest possible score was 6, and Zwicky's grade was 5.45, more than a half-point higher than Einstein had earned before him.

Zwicky often told a story about the oral examination for his teaching degree that some questioned. But the details were so much in character that its essential truth is hard to deny. When one of his examiners asked him a question, the cocky student replied that the question was so vague that he couldn't give a single answer.[14] The second question suited him no better. As observers snickered, the examiner "roared at me" that he would ask his next question straight from a textbook: if Zwicky answered it correctly, he would get a 6; if not, a 0.

A 0 grade would have destroyed his prospects for a diploma, the end of four years of study, and no chance for a teaching job, Zwicky's goal. In his telling, he yelled back that he didn't care what grade he received and walked out. In the end, a faculty patron of Zwicky's interceded with the examiner. "You and Zwicky are both true Swiss," he said. "Why not forget your differences and give Zwicky a grade 5? And that is exactly what happened."

Zwicky became an assistant in the physics department of ETH and settled into what he expected would be a sedate career as an esteemed member of the faculty at one of the world's great universities. In Europe, at least then, the job of professor had a status in society that's hard for Americans to grasp. Zwicky's close friend and colleague in rocketry, Theodore von Kármán, liked to tell a story that illustrated the difference in attitudes between America and Europe: A wealthy American businessman was asked to share the advice he gave his sons about their futures. The first son, he replied, was smart, so he would be a businessman. The second was good-looking, so he would go to West Point. The third "was neither smart nor good-looking so he would be a college professor."[15]

Kármán told that story in late 1957, after the Russians launched Sputnik, an event that shocked and dismayed Americans. To Kármán, the story illustrated the low opinion Americans held about education. That attitude, he believed, went a long way toward explaining the Russian triumph in space.

In Europe, however, a faculty position, at the Institute no less, was the realization of any family's dream. Zwicky was just getting comfortable in his new academic life, studying the structure and behavior of crystals, when a group of American headhunters arrived at the Swiss Federal Institute of Technology. It was August 1925, a relatively quiet interim in Europe between two catastrophic wars and a mere four years before the worldwide Depression would have likely made the headhunters' journey impossible. The heads that Wickliffe Rose and Augustus Trowbridge were after belonged to Europe's educated elite. America was young and wealthy. All it needed, like any member of the nouveau riche, was sophistication. It figured it could buy it.

Rose was a trustee of the Rockefeller Foundation with an impressive résumé in public health. He helped eradicate hookworm disease in the South, greatly improving the lives of thousands of poor African American children.

Trowbridge was a physics professor at Princeton University and a man of wide-ranging interests. He had a particular knack for getting on with, and seeing the value of, people from many cultures. When the two men appeared at ETH one summer's day, they bore more than a million dollars in grants from the foundation. They were looking for the famous mathematician, Peter Debye, a friend of Einstein, to offer him a post in the United States. Debye was out of town, so Zwicky took them around, instead. When he wanted to be charming, there were few who could match Zwicky's talent for robust good humor. By the end of the day, the two Americans were so impressed by their young escort's quick mind and brash confidence that they changed their minds. As Zwicky later told historian R. Cargill Hall, "They asked me, 'how about coming to America?' I asked 'How?' They said they would send me an international fellowship."[16] The fellowship would pay $125 a month. It was no fortune, but enough to get by, if one could adapt to the dry laws. Prohibition, the disastrous experiment in social engineering, was then in its fifth year. Cirrhosis and liver disease declined, but deaths from gunfire were increasing.

Rose asked where he'd like to go in America. Zwicky didn't hesitate. Somewhere "where there are mountains that I can continue climbing."[17]

Rose and Trowbrige looked at each other. Both had the same thought: Mount Wilson.

Forty-five miles northeast of Los Angeles, up the Little Santa Anita Canyon Trail that was so rugged many visitors paid $3 to ride on the back of a mule, Mount Wilson was home to an observatory that was changing forever human understanding of our place in the order of things. There, wrote astronomer Allan Sandage, "would the foundations be laid for modern astrophysics, stellar evolution, the structure of the galaxy, stellar and galactic dynamics, and observational cosmology."[18] Two giant telescopes had been erected, the 60-inch and the 100-inch Hooker. The magnificent Hooker took its name from a local hardware store owner who contributed $45,000 to make the mammoth, 9,000-pound mirror. America had other telescopes, but the stable weather pattern—the inversion layer that later trapped smog in the basin—made southern California the ideal place for the new generation of telescopes being built by George Ellery Hale.

The tortured son of an elevator merchant who got rich rebuilding downtown Chicago after the Great Fire of 1871, Hale had the soul of a gambler inside the delicate constitution of a bookworm. Despite bouts of crippling depression, there were few in any field who dreamed as big, dared as much, and achieved more.

"He who would launch great ships must live in deep waters," Hale is famous for saying. The maxim perfectly described the man who built two of the greatest scientific instruments of the twentieth century, the Hooker telescope on Mount Wilson and, later, the 200-inch Palomar Mountain observatory north of San Diego.

Hale and Zwicky had vastly different personalities. But the rough-edged Zwicky and the wiry, troubled Hale, who talked of being counseled by a churlish demon, would become friends and colleagues in a quest to unlock the universe's most closely guarded secrets. They might not share the same interests: Hale was interested in how stars were born and lived, Zwicky in their violent deaths. But they recognized in each other the same confidence in the value of their explorations.

When Zwicky arrived in California, he had no idea that his future lay in the heavens. Having no training as an astronomer, he came to America as an expert in crystals. The word "physicist" was so alien to American ears that the immigration officer who admitted the young foreigner could not be made to understand what it was. "Physician?" he asked. "Physical culturist" then, he said, doing a deep knee bend to demonstrate.[19]

One of the first people the young Swiss met on his arrival at Pasadena's California Institute of Technology was the head of the Norman Bridge Laboratory of Physics, where Zwicky was to spend his two-year fellowship. By the time the two men met, Robert Andrews Millikan was already a famous man; just two years earlier, he had won the Nobel Prize for determining the charge of an electron. Millikan's experiment was a model of rigor and imagination, two qualities he would demonstrate again and again as the guiding force behind the rise of Caltech from a modest teaching college, Throop University, to one of the world's leading institutes of higher learning. In his experiment, Millikan let oil droplets fall in a glass chamber to measure their velocity, which allowed him to determine their mass. Next, he applied an electric charge to the chamber through which the drops were falling. By adjusting the strength of the electric field, he was able to suspend the droplets in mid-air. This balance between gravity and the electric field enabled him to determine the electric charge on each drop. This allowed him to make a fundamental discovery: the charge on each droplet was always a multiple of a certain basic number. From this he concluded that drops with higher charges must carry more than one electron.[20] It was the first time a subatomic particle had been measured, a major advance in the understanding of the micro-universe.

For someone like Millikan, careful in his speech and manners, a born teacher with wide interests and many influential friends, dealing with a robust mountaineer like Fritz Zwicky was a bracing experience. Many years later, when both men were old and famous, and could afford the convenience of crusty honesty, Millikan admitted that he referred to Zwicky as his "hair shirt."

But Millikan's respect for the younger man's daring intellect was such that he quickly recognized his value and never, throughout their many battles, failed to support him. Millikan was, among other things, a shrewd evaluator of talent; he recruited such scientists and researchers as the physicists Richard Tolman and Paul Epstein, the aerodynamicist Theodore von Kármán, and the chemist Linus Pauling. Under his watch, Caltech began accumulating Nobel Prizes with the regularity that the New

York Yankees won World Series. The latest count stands at thirty-nine, eighteen more than Zwicky's, and Einstein's, alma mater in Zurich.[21]

At the time of their meeting, Caltech had just moved into its new campus across Raymond Avenue, opposite St. Andrew's Church. Pasadena had formerly been known as the Indiana Colony because of all the hard-headed Midwesterners who settled there when Los Angeles was still a rough and rowdy western outpost. In Zwicky's time, Pasadena retained a stubborn conservatism that staunchly favored Prohibition and reliably elected starchy Republicans to all its public offices. Even today, a visitor to the 124-acre campus at the base of the San Gabriel Mountains, with its time-worn stucco-façade buildings named for famous alumni and donors, can't help feeling transported to a time when life moved at a more leisurely pace. In Pasadena—with its tree-lined avenues appointed with frumpy old bungalows and Arts-and-Crafts homes shaded by brick porches against the hot Santa Ana winds that sweep out of the mountains—it still does.

The skies were clear enough that September day to see the green slopes of the mountains north of the city. It was a sight that would become increasingly rare when the automobile imposed its plague of smog on the basin in the 1950s, a scourge against which Zwicky would wage a public campaign. But on this day, he was looking for the rugged mountains, waiting for conquest, which caused him to accept the invitation to California in the first place. Where were they?

Millikan proudly pointed in the direction of 5,700-foot Mount Wilson. Zwicky was surprised. He expected to see something like the bony, sharp-elbowed mountains of Switzerland. Maybe not anything as imposing as the majestic Glärnisch behind his grandparents' house, but something more than the rounded, worn-out old hills slumping off in the distance. They didn't look like they would give an asthmatic octogenarian a good challenge. "Ja," he said. "I see foothills."[22]

It was a disappointing introduction to California. But Zwicky hadn't come all this distance, sacrificing the safety and comfort of his perch in Zurich, to turn around and go home. His fellowship was for two years, and he decided to see what he could make of it. One thing he liked right off was the lack of pretention, especially compared with the rigid, class-based system in Europe. When Millikan introduced him to the university community at a formal affair attended by the faculty and staff of the university, he mixed in enough wit to keep the atmosphere light and engaging. People from the local community were on hand, as well, and all were eager to meet the new arrival. Zwicky found it easy to make friends, falling in with a group of like-minded hikers and outdoors-people with whom he was soon venturing out on trips to Death Valley and Mexico.

He found Pasadena to be a lively place compared to Glarus, although he couldn't get over the fact that southern Californians, even then, drove everywhere. "There is plenty for me to do and everyone here is enthusiastic about everything that is new, and it is addictive," he wrote to his former teacher, Paul Scherrer. "The only thing still needed for absolute perfection is occasional flashes of inspiration."[23]

By this time, he was acutely aware of the competitive nature of his new environment. Maybe they didn't have the credentials and history of the European academy, but they seemed bent on making up for it in their passion and dedication to finding new things. With no history, there were no laurels to rest on, so everything was being made fresh.

"In America you are expected to produce something new every day," Zwicky wrote anxiously to his Swiss friend, Rösli Streiff.[24]

Streiff, like him, was from Glarus. An athletic young woman with a whimsical smile and thick eyebrows who tended to wear her hair in the short bob that was so popular in the Roaring Twenties, she earned a living clerking in her father's bleach factory. Though they hadn't known each other well in Switzerland, their friendship became more devoted the greater the distance between them. Outside his family, Rösli became Zwicky's most reliable connection to his European home. She was a dedicated letter writer, always quick to prop up her friend when he was in danger of losing confidence, and full of unstinting admiration for his accomplishments. Though Rösli never married, and Fritz would marry twice without choosing her, they remained close throughout their lives. By every measure that mattered, and much to her disappointment, their relationship remained chaste.

One of her most attractive qualities, to Zwicky, was that Streiff was every bit as accomplished an outdoors person as he was. Her achievements as a competitive skier brought her fame at a time when it was rare to see

Skiing legend Rösli Streiff, ready to take on the slope. This image was captured in 1932, the year she won the Gold Medal in the Combined and Downhill world championships, in Cortina, Italy. Despite her accomplishments, and symbolic of women's status in the sports world at the time, her father expected her to be at her desk on time in his bleach factory on Monday morning.

women on the slopes. Skiing in Glarnerland has as rich a history as mountain climbing. Ski Club Glarus is the oldest in the country, and Streiff proved herself courageous and agile on the tricky slopes, even in the long skirts required by the times.

She competed in the first alpine world championship ever held, in the Swiss resort of Mürren, 5,000 feet up in the Alps. In 1932, she won the Gold Medal in the Downhill and Combined events, accomplishments that would make her a much-admired celebrity in Switzerland for the rest of her long life. She became something of a matriarch to the younger generation of Swiss alpinists who regularly won Olympic gold. In spite of her successes, her father expected her at her desk on time on Monday morning.

"You probably think we never drink here because of Prohibition and the California sun, but you would be wrong on both counts," Zwicky wrote in an early letter to Rösli.[25] Apparently, the educated class in California, even in solidly conservative Pasadena, paid about as much attention to laws against drinking as their grandchildren would to laws against marijuana use a half-century later.

"There is plenty of alcohol and there has been plenty of rain in the valley and snow in the hills in the last few weeks." So much snow, in fact, that the city of Los Angeles built a ski jump at the Los Angeles County Playground.[26]

"Needless to say, it had to be the biggest in the whole world."

This obsession with the most of everything was the downside in Zwicky's mind to the boundless energy and optimism he found in America. "Always the bigger and better elephant," was the way he put it.

Despite this, he liked his new colleagues at Caltech. "The Americans at the institute with whom I sometimes go skiing pick it up very quickly and are terrible daredevils,"[27] he wrote to Rösli. "But if we include ladies in the event we have difficulties, because most ladies in California have never experienced snow, and they believe they can learn in an hour or two. It's a pity that you can't just come over and show them how to do it."

As dismissive as that sounds, he rose stoutly to the defense of American womanhood when Streiff wrote critically of some American tourists she encountered on the street. "The two 'examples' which you spotted in Glarus must have been especially soft," he wrote. "But you can find that type anywhere. At any rate, most California women are very well built and also quite tough, since they all continually go swimming, play tennis, go to the desert and to the mountains and generally do a very lot of sports, perhaps too much and too ambitiously for European taste."[28]

He couldn't help but put in a dig, as he often did when talking about Americans. "On the other hand, it is very hard for Europeans to get along with American women, since most of them, like their country, are so terribly locked up within themselves and chauvinistically minded."

This was an opinion he would soon find reason to revise.

"As individuals, they of course live, love, and suffer like we do, with the difference perhaps that there are more tall children here than at home."

In the beginning, Zwicky was thought of, and thought of himself, as a specialist in the properties of crystals. This, after all, was his field in Switzerland. One of the first scientific papers he published after his arrival in

California was on the internal makeup of crystals.[29] The primary structure was known to be a lattice arrangement of atoms, but scientists at the time were stumped by the source of cracks that appeared in the structure. Zwicky proposed that rather than a result of weakness, the cracks revealed a secondary structure in the crystals themselves. Millikan, understanding that this idea might lead to stronger materials, was impressed enough to present the theory at a meeting of the American Association for the Advancement of Science, the country's leading scientific organization.

Einstein, when he visited in early 1931, also endorsed Zwicky's idea, which pleased the latter immensely, because, as he told his Swiss friend, Einstein was always "very critical."[30]

Even at this early point, Zwicky showed a remarkable talent, which would contribute to his enduring popularity with the general public, for explaining his ideas in everyday language. One of his first appearances in the American press was an account of an address he gave about his work to a group of local teachers.

"The speaker convulsed the capacity audience of teachers at the outset," reported the *Los Angeles Times,*

> by declaring that atoms "the building stones of solids, lead varying types of private lives." In the gaseous state, the lonely atoms wander about as hermits in a desert and when they shake hands with each other, as they seldom do, they are overcome by their first social debut and vanish.
>
> Atoms in the liquid state have closer neighbors, but continue to act as individual anarchists. This atomic society was likened by the physicist to a congregation of violin players, all of whom fiddle the same tune, but start playing at a different hour. It is only in the crystal state that atoms become civilized, live in perfect harmony and work as a team.[31]

As a layperson's introduction to the atom, one could hardly do better. Zwicky then turned to some behaviors of solids that physicists had been struggling to understand. Glass, he said, was for all intents a liquid. He demonstrated by cutting it with scissors under water. Then there was the fact that it was easier to bend a metal bar than it was to straighten it afterward. Here, Zwicky's idea of the secondary structure of solids came into play. When the bar was bent, the weak atoms loosened their grip. But stronger ones that survived gripped "their fellows for dear life."[32]

Fritz Zwicky was beginning to make a small mark on American science. He certainly had proven to journalists that he was good copy. In succeeding decades, he could hardly utter a thought without local reporters beating a path to his office door. This fact, among other things, would work against him with those colleagues who regarded public popularity as something one should be suspicious of. If the public understood your work, could it be good? If they approved of it, you were in danger of being labeled a panderer and showman.

Fritz Zwicky was without a doubt a showman.[33] Decades later, he would entertain an interested writer when he showed his visitor how he could teach his students without using cumbersome words. To demonstrate, he screwed up his face into one crazed expression after another. Each expression was reproduced in the pages of Caltech's campus magazine. Of course, those who knew him were by then more than familiar with those expressions.

* 4 *

QUANTUM STEAK AND MATRIX SALAD

AS HIS FELLOWSHIP neared its end, Zwicky grew worried. He'd made friends in America and seen something of the country, but he was increasingly anxious over the fact that he was spending his youth with nothing to show for it. Pats on the back from Millikan and praise from local journalists for his glib tongue were nice, but they hardly satisfied his need to do something important. His competitive nature had shown itself in Switzerland, but what he found in America was an entirely different level of ambition. It was a kind of urgent flu that seemed to affect everyone at the institute, an onrushing desperation to bring new things into the world. One couldn't call it unhealthy, because the things being created—fast cars, new electronic devices, high-flying planes—were real, and they were making life easier to manage.

The atmosphere at Caltech also was very attractive. Distinguished scientists from around the world were constantly coming and going. Zwicky had recently met Max Born, the brilliant mathematician who helped formulate quantum mechanics, and the regal, 73-year-old Dutchman, Hendrik Lorentz, whose work formed the basis of Einstein's general and special theories of relativity.[1] Lorentz, chairman of the exclusive International Committee on Intellectual Cooperation, attended a lecture of Zwicky's at the newly opened campus of the University of California at Los Angeles.[2] The subject was, again, the behavior of crystals. Afterward, Lorentz stood and congratulated Zwicky, which thrilled the young man immensely.

One day, Zwicky wrote out an imaginary lunch menu in the whimsical style of the *New Yorker* magazine, which had just begun publication, referencing some of the great scientists he had met and secretly hoped to surpass. "Transparent Soup (F. Zwicky); Quantum steak with mashed Potatoes (P. S. Epstein); Matrix Salad (M. Born); Selection Cake (T. Takamine)."[3] Epstein was Paul Epstein, the esteemed Caltech physicist, known for his contributions to number theory; Takamine referred to Takamine Jōkichi, a Japanese chemist who synthesized the hormone adrenaline. Listing himself among this group showed his need to put himself forward, as well as his hope that there would be a time when he deserved to be listed among those great men. His fondest hope was to win a faculty position at Caltech. But, as the expiration of his two-year fellowship neared in the late spring of 1927, he'd heard nothing from Millikan.

In his darker moments, he wondered whether it would be best to chuck his life in America and return to ETH. There was no guarantee they would take him back, so he wrote and asked his father to be on the lookout for openings on college faculties for an itinerant physics teacher. As he made plans to return by ship and rail after the term ended in June, he couldn't help hoping something might yet happen for him.

But as May turned into June, "the old man," as Millikan was known to the younger faculty (he was 59 at the time), said nothing to him about his future. The graduation ceremony that year took place in a dusty field behind the Faculty Club. The weather was so hot, and it was so difficult to hear Millikan's remarks, that Zwicky left early to pack. His mind was already on his trip home when Tolman and several others came rushing over to congratulate him. "What for?" he asked.

"Well the old man announced that you are a professor of physics," he was told. It was an almost unbelievable gift: Zwicky's career in America had finally started. His drooping confidence burst out anew. True, he hadn't done anything important yet. He would say years later that it took him many years to find his own personal genius. Given that he was just 29 when Millikan rewarded him, this was something of an exaggeration, though he was well aware that great artists and scientists usually had done some of their best and most important work in their early twenties.

His joy was spoiled by tragedy back home. In May, his father had written that his mother, Franziska, was ill with stomach trouble. Now, Fridolin said she could no longer tolerate medicines that might relieve her pain. At first, the Bulgarian doctors thought it was malaria, which was highly unlikely, since she had never been near the tropics. Finally, "the doctors tended to the opinion that Mama was affected in the lungs and therefore recommended a change of climate and mountain air."[4]

He and Franziska decided to return to Switzerland in the hope that the Swiss doctors could do what the Bulgarians could not and diagnose her illness. Going by train along the Danube through Vienna, they arrived back in Glarus, where they consulted a friend of Fritz's, Dr. Fritsche. He was so alarmed he immediately hospitalized Franziska. At the time of Fridolin's writing in late June, she had been in the hospital eight days. She was losing blood and running a fever, but the doctors still could not find the cause of her sickness.

Two weeks later, she was no better. She desperately wished to go to Braunwald, a mountain resort in eastern Switzerland where cars are banned to this day. The cool mountain air, she was sure, would help her breathe better. Dr. Fritsche didn't think she was up to the trip. Fridolin, ever respectful toward professional men, was nonetheless growing frustrated with the failure of the doctors.

"It seems to me that medical science is yet very little advanced," he wrote. The persistence of the fever, he thought, should be a strong hint "to recognize the fog" of the illness, which was likely tuberculosis. Yet the doctors remained stumped.[5]

In late July, Fridolin's report was even more pessimistic. The tone of his letter was despairing; Dr. Fritsche had "ruled out" the possibility of recovery. In his anguish, Fridolin couldn't help striking out at his son. "I especially am concerned that your presence, which had been planned for this summer but was canceled because of your new engagement [meaning the new job] would have helped me get through so much and with your connections to the foremost medical authorities we would have been able to get advice sooner."

This letter could only have rent the heart of his son, to whom his mother was the dearest person on Earth. But the letter, which would have spurred the young man to action, did not reach him before the end arrived.

Franziska Wrcek Zwicky died the night of July 30, 1927. The next morning, Fridolin sat down to write his son a note that, while filled with grief, carried no hint of blame. "She slumbered gently and peacefully away into eternity," he wrote. "I have lost in her a faithful life partner and it will be hard for me without her at my side to wander through my future life."

As the end drew near, he said, Franziska "thought of you especially, asking for you and thinking you near so to speak. May our dear Mama now rest in peace."

The death of his mother was a devastating blow, as it would be to any much-admired son. Busy as he was, with travel and preparations for his new job, he didn't learn until too late the grave nature of his mother's illness. "I still have a deep regret," he wrote years later. "My father was unable to contact me from Switzerland in the summer of 1927 when I was on holiday from my job in America climbing in the Canadian Rocky Mountains."[6] Now he lashed himself with guilt. Had he returned when he promised, he could have intervened and found the best specialists, just as his father had expected. He couldn't blame his father. He was surely competent, but he was too old world and meekly trusting in the expertise and reputations of professional men. Fritz would not have been satisfied with their equivocations and confident assurances that they were doing everything they could. For Zwicky, his mother's death severed one more tie to Switzerland. As he often did when he was troubled, or simply in need of a vacation from his restless mind, he sought relief in the high places.

Despite his first, disappointed, impression of the California mountains,[7] he had wasted little time venturing into the countryside. From the mid-1920s on, he ranged farther and farther afield in search of challenging peaks. He began with the local summits at Mount Baldy, San Gorgonio, and other mountains familiar to southern Californians. One of his first appearances in the American press was not for any scientific work at Caltech but for a failed winter assault on Mount Whitney, just months before his mother's death. By that time, Zwicky had fallen in with a group of friends who regarded mountaineering with the same passion and devotion to hardship that he did.

"After you've done it for a while, mountain climbing grows to be almost a religion," Kem Weber told the *Los Angeles Times*.[8] Weber, a furniture designer and a leading exponent of the new West Coast Aesthetic emphasizing casual elegance, was one of Zwicky's first German expat friends. The group became known around Pasadena as the German Club; they held dinner parties that became more and more elaborate as time went on, featuring exotic courses at a time when there was only one Swiss restaurant in the whole of the Los Angeles basin.

Whitney, at 14,500 feet, is the highest peak in the lower forty-eight states. About 220 miles northeast of Los Angeles, the mountain is the result of a fault system on the eastern side of the Sierra Nevada Mountains that even today slowly pushes the mountain higher, a bit like someone raising a cellar trap door. In the summer, the trail to the granite peak, high above the tree line, is relatively easy; the hardest part is getting used to the shortage of oxygen at altitude. It's customary for outdoors-people these days—and there are so many they can resemble a marching band more than a troop of hikers—to camp overnight to acclimate. In winter, it's a different story. There were no records until that time that anyone had summited Whitney when the mountain was deep in snow.

Zwicky, Weber, and a third man, unnamed, set out on Sunday, March 27, each carrying a knapsack weighing fifty pounds, containing an ice ax, crampons, and a tent made of "balloon silk." Their trek started at 3 A.M., in Lone Pine Canyon, and after three hours they reached Mirror Lake, 10,500 feet up. There they made camp and rested most of the day before renewing their assault at 11 P.M., climbing through Whitney Pass at 13,500 feet by 4:30 A.M. Monday.

It was so cold that they stopped and had some soup before trudging on. They were less than a thousand feet from the top, but the going was slow and laborious. They had mistakenly taken a particularly difficult, roundabout route, consuming three hours and reserves of energy. "Every step had to be hewn out with an ice ax, and a single false move would have meant disaster," Weber told the *Times*. "But at length they reached a point 13,800 feet high . . . and closer, Weber believes, than anyone has ever been under the circumstances."

Their last stores of energy were used up within sight of the summit. "The altitude and the cold had so sapped their energies that their progress," Weber said, "was like the action in an ultra-rapid motion picture. They all wondered why they had ever had the folly to attempt such a climb."

It was interesting that Zwicky was not quoted. Given his attitude about the poor quality of the mountains in California, compared to his beloved Alps, it would have been revealing to know whether Whitney had changed his opinion.[9] As for Weber's fevered description, it must be remembered that there were no trails to the peak at the time and that their equipment was so deficient that it made a winter climb of any sort difficult. A picture accompanying the *Los Angeles Times* piece, which covered almost an entire page (evidence of how important the editors considered the attempt) showed Zwicky trudging up a nearly vertical bank of snow, wearing a cloth alpine hat and heavy coat. A canvas bag was slung over his shoulder on a strap. For support, he used a simple walking stick.

His appreciation for the mountains in the New World continued to grow as he journeyed farther afield. After climbing 9,760-foot-tall Mount Robson, which he referred to as the "monarch of the Canadian Rockies," he gushed to Rösli Streiff about the grizzly bears, moose, and other wildlife he glimpsed on the way up. When some colleagues were stranded in the snow in the Saline Valley in eastern California, several years later, it was Zwicky who led a rescue party.[10]

He had little good to say about the equipment he found in the United States and endeavored to improve on it, attracting attention for the first time outside the southern California outdoors community and local science journalists.

"It was Dr. Zwicky and some of his colleagues in Zurich who introduced short broad skis (summer skis) for mountain climbing," *Scientific American* said, in a short piece that introduced the Swiss crystals expert to a wide audience.[11]

"He also improved the technique of climbing steep ice slopes with crampons. Slopes up to 60 degrees may be climbed without chopping steps, by backing up."

Other writers treated this innovation—going backward up a mountain!—in the same not-quite-breathless but still admiring tone. When he showed that one could as easily ski on sand dunes in Death Valley as on the slopes of Mount Baldy, that was celebrated, too.

The most challenging and dangerous of his climbs took place in 1930, not in North America but back in Europe, in the Pennine Alps, on the border between Switzerland and Italy. A climbing partner and friend,

Max Brunner, provided a harrowing description of the adventure that Zwicky saved for the rest of his life.[12] The mountains opposing him were Dent d'Hérens, a 13,684-foot peak near the Matterhorn, and Dent Blanche, 14,295 feet high.

"My companion was Dr. F. Zwicky," Brunner wrote, "a Swiss, whom I knew since long [*sic*] and who is a most expert Alpiniste."

They began the Hérens climb at the Schönbühl hut nearly 9,000 feet up in the Alps. The hut keeper warned them that six parties with guides had failed to make the ascent because they could not pass a dangerous icefall. Zwicky insisted on making the attempt.

They got through the first icefall, confronting a beautiful sight of "thin ice bridges climbing up and down into blue crevasses." At the second icefall, they spent two hours cutting steps into a wall of ice, only to be stopped by a huge crevasse that was impossible to cross. They turned back and made the attempt over loose rock that finally brought them to the top of the second icefall. Finally, they reached the northwest face of Hérens, where they spent four hours cutting ice steps until they reached "a very steep wall of pure ice, about 100 feet high, with an angle of sixty degrees." It took another hour to overcome this barrier, which involved not only cutting steps but also handholds in the ice.

"Here especially we had to trust each other because one's slip would have necessarily involved the destruction of both," Brunner wrote.

That problem was solved only to yield to another, an enormous crevasse. "The other side of this crevasse consisted of an ice overhang of fifty feet and in vain we tried to find a weak point to take him," Brunner wrote. Crevasses were a constant danger to professional as well as recreational climbers. On a later climb, Zwicky and his partner would both tumble into separate crevasses at the same time, dangling at the ends of their ropes, each waiting for the other to pull him up. The event was the subject of a cartoon that made the rounds at Caltech. It was amusing only in retrospect.

On Hérens, they couldn't get over this last hurdle and, just as on Whitney three years earlier, they had to turn back within sight of the summit. The sun set as the climbers laboriously made their way down the two icefalls. At 10 P.M., after nineteen hours of labor, they reached Schönbühl hut.

Zwicky the mountaineer at Zermatt in 1932. The alpine region of the Pennine Alps, in the canton of Valais, attracts climbers from around the world. In the background is 11,400-foot Col de Zinal.

They spent a day recuperating, when they were joined by Tadeusz Reichstein. Zwicky was so delighted by the appearance of his old friend that they began the assault on Dent Blanche as early as possible, at 2 A.M., the next day. The difficulty this time was the treacherous rock face. The chimneys were steep, and the rock was loose. A two-pound stone dislodged by one of the three struck Brunner in the head.

"For a short moment I was nearly fainting and I fell some inches," he wrote. Luckily, all three climbers were roped together, which saved him from a worse injury, if not death. The stone left a deep cut on the man's head, which gushed blood. "The blood was covering my whole face and windcoat and was streaming with such a speed that I thought I had to loose [*sic*] all my blood."

After a few minutes, the injury crusted over in the alpine chill, and the party decided to keep going. The pace was slow, however, and after sixteen hours of climbing they halted for the night. Mountaineers in Zwicky's time had none of the modern nylon tents, goose-down sleeping bags, or sturdy hammocks that one could fix to the rock wall and pass the night in some degree of comfort. Their only provisions were candles, lanterns, some chocolate, and newspaper, which they stuffed into their boots and around their ankles to prevent frostbite. Due to the precarious nature of the bivouac, only one man at a time could lie down. The others had to sit upright. A fog bank descended, and a thunderstorm over the Matterhorn electrified the night sky, a wondrous, and terrifying, demonstration of nature's power. "Never in my life I saw so many lightenings [*sic*]. During the whole night we were looking anxiously at the weather, for it was certain to each of us that a change of the weather to the worse would be most serious."

In this way, the expedition spent the night, taking turns lying down, anxiously debating their options if the storm swept over them and buried them in snow. There were no real choices. They could not go up or down in the dark. They would simply be entombed. The next morning, stiff and dazed from lack of sleep, they rose and resumed their trek. At noon, they reached the summit, where they met another climbing party. The strangers gave the exhausted companions bacon and bread. That night, the party stumbled into the Schönbühl hut, Brunner scaring the occupants with his blood-caked face.

The three discussed the possibility of next trying the Matterhorn by the challenging Zmutt Ridge. Reichstein and Brunner were uncertain. For one of the few times in his life, Zwicky agreed to retreat. Having absorbed their full shares of punishment, they hiked down and took the train from Valais to Zurich.

What was the purpose of all this sweat and danger? The mountains were indifferent to a person's exertions. Zwicky addressed the question in his book, *Jeder Eine Genie* (Everyone a Genius).

"Non-alpinists keep asking why we keep madly racing up mountains," Zwicky wrote. Many answers were possible, "the wonder of nature, physical workout, escape from daily life or simply the joy of the adventure. The German poet Schiller said, *'Und setzest Du nicht das Leben ein, nie wird Dir das Leben gewonnen sein'* (Unless you stake your life you can never gain it.)."

Zwicky had a different answer. In science, "there are hardly any problems which you can solve completely, on your own. . . . Even if you have success initially, you become aware of fresh aspects which occupy you for a long time, perhaps for the rest of your life. That is why we yearn for examples of success which can be completed, something we have done alone, something no one can take from us."[13]

Though he would achieve great things, explain behaviors that had been hidden, the deepest truths of the cosmos always seemed just out of reach. And of course, there was the feeling that the things he did find were stolen from him by men with lesser imaginations. Sometimes this suspicion was justified; at other times, it was evidence of an unflattering streak of distrust and insecurity that would drive away friends and allies alike. Many years later, the author Bill Bryson would sneer at Zwicky's efforts to prove his physical prowess. In his book *A Short History of Nearly Everything*, Bryson, reflecting the opinion of Zwicky's detractors, said Fritz "would often drop to the floor of the Caltech dining hall or other public areas and do one-arm pushups to demonstrate his virility to anyone who seemed inclined to doubt it."[14]

Whatever the tensions in the scientific world, no one could take from him that awful, exhilarating conquest of Dent Blanche in the summer of 1930.[15]

By this time, Zwicky was thirty-two, and had yet to make any lasting contributions to science. All that was about to change, as a result of a jab from Millikan. Actually, it was Zwicky who first jabbed Millikan. "Mr. Millikan," Zwicky said, in a story he recounted often, "I have read every paper you ever wrote, I have listened to every presentation you have ever given, and I can tell you quite categorically that I have never found a single original idea that you could honestly call your own."

Typical, risking all on a bold assault. It was also impolitic for an associate professor. But Millikan was a wily supervisor, used to dealing with cocky young men trying to make their marks on the world. He didn't give the back of his hand to his young colleague, though he must have been tempted. Instead, he turned the younger man's weapon against him, replying, "All right, how about you?"

"I have an original idea every two years," Zwicky replied. "I'll go further: you name the subject, I'll come up with the new idea."

"All right young man," Millikan said. "Astrophysics."

* 5 *

THE EXPANDING UNIVERSE
AND TIRED LIGHT

IN 1785, THE BRITISH composer-turned-astronomer
William Herschel developed one of the first theories of cosmogony, or
the origin of the universe. Herschel was a very fine oboist, who counted
Mozart and Beethoven among his fans. He also discovered the planet
Uranus. Both achievements were demonstrations of great insight and
steady perseverance. Yet his ideas about how the universe was arranged
were understandably limited, given that the telescopes he built to study
the heavens used metal discs of tin and copper for mirrors. On the basis
of his research with these rudimentary tools, he believed that space
was seeded with stars in an orderly fashion, as if by plan.

By the late nineteenth century, telescopes, and the imaginations of researchers, had grown large enough that astronomers launched themselves
on a daring effort to map the continents of space in a more careful way.
The *Carte du Ciel* (chart of the sky) was a breathtakingly ambitious, and
fatally flawed, plan to use the world's observatories to identify the positions and brightness of millions of stars brighter than the 12th magnitude
(the faintest star visible to the human eye is about 6th magnitude). Tens
of thousands of stars were collected on 22,000 glass photographic plates.
But the effort failed; there were simply too many stars and too few eyes to
count them all.

Jacobus Kapteyn, a Dutch astronomer born a century after Herschel,
in 1851, had a better idea. He thought that by carefully surveying 683

representative areas of the sky,[1] the effort could be made manageable and might perhaps uncover clues to "the formation or evolution of the 'universe,'" in the words of astronomer Allan Sandage.[2]

Crucially, his idea caught the fancy of George Ellery Hale, who had just finished work on the first great telescope on Mount Wilson, the 60-inch reflector. Hale adopted Kapteyn's idea as the first important research program at the observatory. Kapteyn's findings, "First Attempt at a Theory of the Arrangement and Motion of the Sidereal System," was published in the May 1922 issue of the *Astrophysical Journal*. From his work, he drew several conclusions, one of which—that the Milky Way was slowly spinning in space—was inspired and more importantly correct. The Kapteyn Universe, as it came to be known, was a flattened disc, a mere 32,000 light-years across.

He also thought he saw evidence that our solar system was surrounded by two streams of stars,[3] swirling in opposite directions, like great rivers rushing to separate seas. This was woefully wrong, but, unfortunately for Kapteyn, the scientific establishment latched onto this idea with relish, as people tend to do with a good idea that seems to bring order to any apparently random system.

Kapteyn's great error stemmed from the fact that he and his observers were misled by a forest-and-trees problem, or, in Sandage's words, a beehive-and-bees problem. From the standpoint of a tree, individual bees are flying around the hive on various pathways, in three main directions: toward the hive, to the left, and to the right. As Karl Schwarschild later proved, in an early demonstration of stellar dynamics, the stars only appear to be moving in streams, when in fact the motion is much more random. Sadly for Kapteyn, whose work was good and groundbreaking, the idea of star streams was so attractive that he held fast to it to the end of his days, which arrived in 1922.

Despite these early efforts to understand the makeup and arrangement of the heavens, it was still possible at the dawn of the Jazz Age for the average American to believe that the universe was organized, in all its finery, simply to entertain God's great creation, humankind.[4] At the center of the stage was the sun.[5] Discoveries at Mount Wilson and elsewhere, however, were challenging that idea. Harlow Shapley, a former newspaper crime reporter from Kentucky, who was one of the leading sky surveyors on the mountain, had been making measurements of star clusters that he said proved the sun was nowhere near the center of things. Earth, he said, was in fact two-thirds of the way toward the edge of the Milky Way, a mere exurb in the cosmos.

Up at the Lick Observatory in northern California, Heber Doust Curtis, once described as a "small, quiet man with a remarkable sneeze," had been making photographs of blurry objects called spiral nebulae, a term lifted from the Greek, meaning cloud. Rather than drifting blobs of interstellar gas, nebulae were, Curtis believed, entirely separate galaxies, so far away that telescopes could not resolve them into stars. For Curtis, the Milky Way was just one of who knew how many galaxies in the cosmos. This was known as the island universe theory. Studying these islands, figuring out how they formed and what held them together, would be where Fritz Zwicky would make his mark in the young field of astrophysics. But at the time, the idea of a cosmos composed of billions of galaxies, each containing billions of stars, was beyond the reach of most people's imaginations.

If true, these discoveries by Shapley and Curtis would shrink the solar system to the status of a one-stoplight hamlet. Even though both men were on the same side in challenging the idea of a sun-centric star field, they disagreed about enough other things that they became competitors in a contest that had scientists around the country, and the world, choosing sides. To Shapley, if the nebulae were other island universes, then the Milky Way "was a continent." For his part, Curtis sneered at Shapley's assertion that the sun was out on the fringes of the galaxy. The argument wasn't so much about the evidence. Both men had access to the same images of the sky. The question was what those images were, drifting clouds of gas or galaxies like our own. Of the two, Curtis was the more imaginative, envisioning a far grander universe.

The dispute reached such a pitch that George Hale, a leading member of the National Academy of Sciences, arranged to have Shapley and Curtis confront each other in Washington, DC, on the night of April 26, 1920. The title of the conversation as it was called was, immodestly enough, "The Distance Scale of the Universe."[6]

Hale had originally chosen relativity as the topic, but C. G. Abbot, the home secretary of the Academy, objected. "As to relativity, I must confess that I would rather have a subject in which there would be a half dozen members of the Academy competent enough to understand at least a few words of what the speakers were saying," he wrote back. Like many scientists at the time, who weren't convinced that Einstein's theory of gravitation was right, or that it mattered even if it was, Abbot was sick of the subject. "I pray to God that the progress of science will send relativity to some region of space beyond the fourth dimension, from whence it may never return to plague us."[7]

So, by default, the discussion centered on the island universe theory. History would record it as the "Great Debate." While on its surface the problem seems like a mere statistical difference, an argument over the arrangement of crockery in the cupboard of the universe, greater stakes were involved. The argument aimed to settle nothing less than the question of humanity's place in the universe. When it was over, and after the dust had finally settled, human beings would find themselves in a cosmos so wild and obstreperous that many doubted it could be the product of divine design. It was a cosmos that had to come into being, however, to contain the monsters that Dr. Zwicky would show to the world only a decade later.

A large, formally dressed crowd of scientists from a variety of disciplines turned out for the afternoon discussion at the Smithsonian Museum of Natural History. Curtis and Shapley were not Lincoln and Douglas. Shapley was so timid a public speaker that he read his speech from typewritten pages, gripped tightly in sweaty hands. Of his nineteen pages of remarks, he meandered for six before he got to the definition of a light-year. Curtis, with a background as a lecturer in Latin and Greek, was by far the more self-possessed. But his talk was so technical that few in the audience could understand it.

Still, Curtis was happy with his performance. Afterward, he wrote home that "the debate went off fine in Washington, and I have been assured that I came out considerably in front."[8]

As it turned out, Curtis prevailed not only on the debate stage but in fact. Three years later, Shapley's colleague at Mount Wilson, Edwin Hubble, proved that the Andromeda Nebula was in fact another galaxy.[9] Shapley took the news badly. The astronomer Cecilia Helena Payne (better known now by her married name, Cecilia Payne-Gaposchkin) was in Shapley's office when a letter arrived containing the final proof that he was wrong. Waving it in the air like a warrant for his own arrest, he is said to have griped, "Here is the letter that destroyed my universe."[10]

His universe had to be destroyed to allow a far grander and more magnificent one to rise in its place. His measurements at Mount Wilson placing Earth on the far edge of the Milky Way turned out to be correct, however, validating both his, and Mount Wilson's, reputations for excellence. As a result, he won appointment to the directorship of the Harvard College Observatory, one of the most prestigious posts in astronomy.

Fritz Zwicky was not at first on the staff of Mount Wilson, but he visited the observatory and was well aware of the discoveries coming out of there that were mapping the expanse beyond Earth. His first astrophysics paper, "On the Thermodynamic Equilibrium in the Universe,"[11] was typical of him. Rather than put a toe in the water of a new field, he jumped into the deep end of the pool. The paper challenged the idea that the Second Law of Thermodynamics would cause the universe to run down. According to that law, entropy or disorder always increases. Applied to the universe, the idea suggested that eventually, in some distant time, all the fuel in the stars would be used up and the light would go out, a process known as "heat death," leaving a frozen Earth to drift along the darkened hallways between dead stars.

In his paper, Zwicky admitted that it "seemed to be justified" at first glance that the billions of stars would run down "irrevocably toward a state of highest entropy." That would mean "all the matter would disappear into radiation." After reducing the constituents of the universe—radiation, dust, stars, etc.—into a set of equations, he said the idea that there was an equilibrium in the universe did not appear to be "justified by the facts as far as the distribution of radiation in the universe and the equilibrium between matter and radiation is concerned."

That didn't mean the theory was wrong. His view was more nuanced. Only when one considered such facts as the "perfect" regularity of "agglomerations of mass" in star systems and the interaction of vast quantities of dust and vapor, could one decide whether the universe was, or was not, running down.

Because there simply wasn't enough information, the answer could not be found with the scientific tools available, he said. His preferred answer was clear, nonetheless. He was constitutionally unable to believe that in the struggle between order and chaos, disorder would prevail. He applied the same idea to human relations. No matter how bad things got, he felt, great individuals would always rise to straighten things out. It was a very romantic, and old, idea, the belief that the arc of history leaned toward justice. It would be tested to the breaking point by World War II.

This paper alerted the scientific world to the arrival of a fresh voice in the budding field of cosmology. In one of his first appearances in the national media, *Fortune* magazine referred to him as "alpine mountaineer, colleague of Einstein, and one of the most brilliant theoretical physicists at Caltech."[12]

At the time, Zwicky was doing a lot of thinking about cosmic rays, the subject that had been bedeviling the scientific world for several years. The furor had reached such a pitch that Zwicky even addressed it in a note to Rösli Streiff, saying that his colleagues at Caltech were "in great agitation about cosmic radiation."[13]

So much so that the famed British cosmologist, Sir James Jeans, decided to visit Caltech and consult with Millikan's "noted young men of science." Sir James was a popular author and a leading advocate for what came to be known as the steady state theory, which held that matter was being continuously created throughout the universe.[14] That theory would eventually be discarded, but at the time, he was one of the Western world's most influential thinkers.

When it came to cosmic rays, he and Millikan were on opposite sides. As one newspaper characterized it, "Is a cosmic ray a birth cry, a death wail, or just a penetrating yell?"[15] Millikan believed the first, Jeans the second possibility.

"It may be we are both wrong but I look forward to discussing the matter," Jeans said. At a lecture at Caltech's faculty club, the Athenaeum, Jeans told scores of worried listeners that there was no reason for alarm over what was happening in the universe.

He also conferred over lunch with Zwicky about some measurements Edwin Hubble had been making at Mount Wilson, measurements that were about to upend every idea about the cosmos. Sir James said Zwicky had a theory about Hubble's work "that strikes me as very explanatory."[16]

Meanwhile, in a letter to *Physical Review*,[17] Zwicky was still tussling with the question of whether cosmic rays were produced locally. Privately, however, he had begun to think more imaginatively. In a lecture at Caltech covered by the *Los Angeles Times,* he said that some, at least, must originate outside the Milky Way. But he teased that he was not yet in a position to say where exactly these intergalactic emanations came from. That answer would await the Stanford University conference.

Although Mount Wilson had rightly earned a reputation as the leading observatory in the world, George Hale was already looking ahead. As the decade of the twenties passed, and the size of the universe grew larger and larger, he realized that the 500-million-light-year reach of the 100-inch Hooker telescope would not be enough to penetrate the vastness of the cosmos. In 1928, he convinced the International Education Board of the Rockefeller Foundation to set aside $6 million to build an even bigger telescope. The 200-inch "Big Eye," one of the greatest engineering projects since the pyramids of Egypt, was given directly to Caltech. That meant the university would be charged with managing the great project, with letting contracts to grind the 14.5-ton mirror and construct the 135-foot-tall dome, a height chosen to match the Pantheon in ancient Rome. The Hale telescope, when it was completed, would gather four times as much light as the Hooker and allow observers to see a billion light-years into space. Those observers must be trained, and that's surely what Millikan had in mind when he challenged Fritz Zwicky to enter the field of astrophysics.

Once Hale had managed to wrest the money from the Rockefeller Foundation, the question became where to put the new instrument. The discoveries on Mount Wilson had proved the superior characteristics of southern California weather. The air was still and the nights crystal clear. The growing population in the Los Angeles basin, however, with its attendant light pollution, showed that Mount Wilson would no longer be suitable. Its day was already passing. Palomar Mountain, a 6,100-foot-tall peak northeast of San Diego, was suggested as a possibility. Clothed in a thick forest of oak, pine, and cedar trees, the mountain was known to the local Luiseño Indians as Paauw; it was the Spanish who named it Palomar, which meant "Pigeon Roost," for the band-tailed birds that flocked there.

Zwicky was part of the first scientific team to reconnoiter Palomar, which at the time was a trackless cattle range. Milton Humason, a former mule-driver who had become a vital part of the Mount Wilson observing team, was hiking back to the ranch house, where the team was bunking during a scouting trip, when an irate cattleman pulled up in a Model T truck. "So you're the cockeyed stargazers been scarin' my steers," the man said, ordering the group to hop in back. After driving them back to the ranch house, he told the landlady to keep the scientists locked up. "Every time you scare a bunch of them steers I lose a hundred pounds of beef. You fellas was costin' me money."[18]

The rancher's warnings went unheeded. Zwicky found a spot near the peak and set up a small, portable telescope. The air was clear and still. Several small Indian reservations spread out below, but they were simple, humble settlements. Decades later, those reservations would become home to casinos, bringing light that would interfere with Palomar's great telescope. But in 1930, the skies were clear and dark all the way to the Pacific Ocean, fifty miles to the west. On the basis of the team's report, the site was chosen and work begun.

Hale would not live to see his great instrument completed. A true marvel of twentieth-century engineering, "first light" would be delayed by the second world war, until 1948. Despite what Hale called his "head trouble," he never lost his enthusiasm for the tapestry that spread over that troubled head each night. Shortly before his death in 1938, he declared it a "beautiful day. The sun is shining and they are working on Palomar."[19]

As Zwicky was turning his attention to the heavens, he received a tantalizing inquiry. His alma mater, the Swiss Federal Institute of Technology, had come forward with a job offer: the assistant professorship of Theoretical Physics. It was a prestigious post formerly held by Einstein and Erwin Schrödinger. Schrödinger is popularly known today for the thought experiment known as Schrödinger's Cat, which addresses some of the most confounding elements of physics, suggesting that, so far as the science at the tiniest dimensions are concerned, a cat in a sealed box could be both alive and dead at the same time. Within science, he is better known for the famous equation that is the foundation stone of quantum mechanics, "the starting point," in the words of physicist Nazim Bouatta, "for every quantum mechanical system we want to describe: electrons, protons, neutrons, whatever."[20]

Zwicky may have led his family to believe he still thought about returning home, but when he had the chance, he made demands that the Zurich institution found impossible to meet. The money would not suit, he said. If this was a bargaining chip, it failed. The faculty selection committee did not even counter, instead expressing disappointment and hoping he might reconsider later, when "he had made his name in America and would be a credit to the Swiss academic world."[21] By this, the committee meant when he became famous enough to justify the salary he demanded. It was clear from the negotiations that the administrators at the Swiss institute had serious doubts that he would ever be worth the money. They would be wrong.

One must admire Zwicky's chutzpah in his choice of targets. First, he had taken on Millikan. Next, he challenged the conclusions derived from the work of Edwin Hubble, considered by many to be the most influential astronomer since Copernicus. Hubble sat atop Mount Wilson as if it were Olympus. A balding, touchy, reticent man who always had a pipe projecting from his mouth—astronomers of that era were inordinately fond of that particular affectation—he had been a natural athlete as a boy in Chicago. At 6 foot 3, he was attracted to the basketball court, but he always felt his true calling rested much higher than a metal hoop. He became a Rhodes Scholar and taught high school before deciding, at the relatively late age of twenty-five, to follow his heart and become an astronomer.

Hubble arrived at Mount Wilson in 1919, six years before Zwicky landed at Caltech, becoming part of the greatest assemblage of astronomers in the twentieth century. Besides Harlow Shapley, the observatory would be home to Walter Baade, who doubled the size of the universe; Milton Humason, who helped discover that the universe was expanding; Rudolph Minkowski, who discovered there were two classes of supernovae; and Allan Sandage, Ralph Wilson, Fred Seares, and Horace Babcock, among others.

Mount Wilson was a cliquish club. They all knew they were doing important work and strove to outdo one another. By the time Fritz arrived in Pasadena, Hubble was already a famous man. He first achieved renown by discovering a star known as a Cepheid in the Andromeda Nebula,

Edwin Hubble was a towering figure, both physically, at 6 foot 3, and in terms of his monumental discoveries. He helped prove that the universe is expanding, which Zwicky could not accept.

proving that it was indeed a separate galaxy, the evidence that destroyed Harlow Shapley's universe. A Cepheid is an unusual star that brightens and dims at regular intervals. The star discovered by Hubble, known as V1, for variable star number one, has been called the most important star in the history of cosmology, because it allowed astronomers for the first time to measure vast galactic distances.

Simply measuring the apparent brightness of a star doesn't tell you how far away it is. The most luminous stars may appear to be dim if they are very far away, or if there is a lot of dust in between them and us. Henrietta Leavitt, of the Harvard College Observatory, discovered that the intrinsically most luminous variable stars had a longer period between their peaks of illumination than their dimmer brethren. Astronomers suddenly had a good yardstick to measure vast distances in space.

Using Leavitt's discovery, Hubble found that the Andromeda nebula was a million light-years away. And just like that, the size of the universe suddenly "blew up," as science journalists of the time described it, beginning its march, over the coming decades, to ever greater and more astounding sizes. Until today, when scientists feel they have a pretty good handle on something close to a final size: 92 billion light-years in diameter, and around 13.8 billion years old.

But Hubble's greatest discovery came a few years later, in 1929, as America and the world entered the greatest economic calamity of the century. The question at the heart of his discovery was: what kind of a universe do we live in? Was it the static, stable, reliable place that many scientists believed in? A place of infinite beauty and reliability, a place that a rational God would have in mind when he set about building it? Evidence was growing that there was something wrong with this idea.

Georges Lemaître, a Belgian priest and cosmologist, argued that a static universe of constant size would be inherently unstable.[22] The slightest movement, a sneeze, a butterfly flapping its wings, could set off a chain of events sending the universe into a frenzy. He introduced the idea of a universe billowing outward in all directions, very much like a balloon being inflated. But he had no proof.

As early as 1912, Vesto Melvin Slipher, of the Lowell Observatory in Arizona, had begun analyzing the light coming from far-off nebulae by breaking down the light rays in a spectrograph. This produced a fingerprint of the light source, basically a series of lines representing the absorption or emission of light in a narrow frequency range.[23] Slipher discovered that most of the galaxies exhibited redshifts, meaning the light was shifted more toward the red end of the spectrum.

A simple way of thinking about this is to compare the process to the Doppler effect, which we experience in everyday life in the way an ambulance siren rises in pitch as it gets closer and drops in pitch as it moves away. The reason is that the speed of the approaching ambulance is added to the sound waves moving ahead of it, compressing them and shortening their wavelengths to produce a higher pitch. As the ambulance speeds away, the wavelengths are stretched, producing a lower note. With light, this "lower note" means a redder, longer light wave. The faster the wave is traveling away, the redder the light. The comparison between light and sound is not exact, because sound travels through the air and light travels through a vacuum. Another factor involved in the redshift puzzle was the fact that, as Einstein proved in his general relativity equations, the waves of light are being stretched by the expansion of space-time itself. It is not so much that the galaxies are speeding away, but that space itself is billowing out in all directions like Lemaitre's inflating balloon, carrying everything with it.

Slipher found that some other galaxies had larger redshifts than stars in our galaxy. This meant the stars in other galaxies were traveling away from our region of space at faster speeds. There was a deep mystery at work, Slipher realized. Were there tides in the cosmos? Was there a great wind raging in deep space, sweeping everything away?

It took Hubble, and Humason, working at the 100-inch Hooker telescope, to prove that Slipher—and Lemaître—were onto something very big. Training the telescope on far-off star systems, Hubble found greater and greater speeds, eventually discovering that speed and distance were related. If a galaxy had ten times the redshift of another star system, it must be ten times as far away and was traveling ten times as fast. This is now known as Hubble's Law. These were the strange measurements Sir James discussed with Zwicky over lunch.

Hubble's discovery landed in popular culture like an exploding bomb. It challenged everything astronomers thought they knew about the universe. Forget Curtis and Shapley and their arguments about the sun's place in the cosmos, Hubble's discovery dynamited the entire structure of the universe. It was, in Allan Sandage's words, "the most fundamental discovery in cosmology for all time."[24]

It was also an idea that Fritz Zwicky could not accept, sneering at those who believed "all of the blah blahs of the expanding universe."[25] As the measurements of star systems came in with ever greater speeds attached, he grew more skeptical. In 1931, Humason came up with a velocity of 19,600 kilometers per second (12,179 miles per second) for a galaxy cluster in the constellation Leo. These enormous speeds convinced Zwicky that the theory was dead wrong. He was hardly alone at the time. James Jeans dismissed the breathtaking speeds being found as impossible. Zwicky immediately set about trying to figure out why the measurements being made by Hubble and Humason didn't mean what they appeared to mean: that the universe was blowing up.

It was known by this time that gravity could bend light. Einstein had shown as much with his general theory of relativity. Sir Arthur Eddington's famous 1919 observation of a solar eclipse, when he measured the exact amount of bending from stars in the Hyades cluster as the light passed by the sun, was the event that transformed Einstein into a pop culture superstar.[26] Zwicky now used this idea to argue that the extreme redshifts Hubble was finding were not caused by high recessional velocities. Instead, they resulted from the fact that, over long cosmological distances, photons of light pass so many massive objects that they lose some of their energy. He elaborated on the idea in a paper for the journal *Helvetica Physica Acta*. As a photon heads toward Earth, he said, "it loses on the way from P_1 to P_2 a certain momentum and gives this to matter. The photon gets redder."[27]

This came to be known as the tired light theory. It received widespread attention at the time. Referring to him as the "the colorful Dr. Zwicky," whose name "kept coming up in discussions amongst the most eminent scientists in cosmology," *Scientific American* wrote that if "this theory is correct, we can presumably return to Einstein's earlier hypothesis of a finite but unlimited universe." Hubble himself was impressed by Zwicky's idea.[28] He was a cautious man, after all. Even he found his results hard to believe.[29]

In the preface to his book, *The Realm of the Nebulae*,[30] Hubble went out of his way to thank Zwicky for his contributions to Hubble's understanding of the universe. "In the field of cosmology," he wrote, "the writer has had the privilege of consulting Richard Tolman and Fritz Zwicky of the California Institute of Technology. Daily contact with these men," who also included Humason, Baade, and Sinclair Smith, a pioneer of dark matter research, "has engendered a common atmosphere in which ideas develop that cannot always be assigned to particular sources. The individual, in a sense, speaks for the group." It was clear from this and other things he said, that Hubble's caution was deeply influenced by Zwicky's skepticism.[31]

Even while putting forth his own theory, Zwicky admitted that none of the proposed theories to explain the redshift "is satisfying. All of them have been developed on a most hypothetical basis, and none of them has succeeded to uncover any new physical relationships."[32]

For one thing, if the light from remote objects had been interfered with on its way to Earth, the effect should show up in the telescope. The object should be ever so slightly blurred. This blurring was not seen.

Of course, he had to admit that with the current technology, any results were provisional. Making photographs of distant galaxies, even with the superior technology of the 100-inch Hooker telescope, required such long exposure times that the results became problematic.

Nonetheless, more and more proof accumulated. Finally, a consensus was reached. The universe was indeed expanding, and at a frantic pace. Everything we see around us in space is in a desperate, headlong race to oblivion. It was only a matter of time before the cause was found, the starting gun when the race began: the Big Bang.[33]

Today, the tired light theory has been sent off to the fringes of astrophysics, where crackpots argue over time travel and alien visitations. Zwicky was no crackpot, but he could never accept that the universe was "blowing up." Throughout the rest of his life, he wrestled with the idea of an expanding universe like a man struggling with a demonic force.[34]

Some might argue—and some did—that Zwicky's refusal to budge on one of the most settled tenets of modern cosmology revealed a deep flaw. To his critics, this characteristic exposed him as a common contrarian, a crank with a knack for making lucky guesses about supernovae. To them, his failure to accept the obvious was a fatal shortcoming.

To his advocates, and the number of them continues to grow as Zwicky's many feuds are forgotten, this stubborn insistence on his own truth was an important part of his inspirational process. "There is no question in my mind that Zwicky was a greater intellect than Hubble," said the astrophysicist, Richard Ellis, whose work draws on Zwicky's insights.[35]

For Caltech's George Djorgovski, another fan, Zwicky's eccentricities put him in the same category as Johannes Kepler, the great seventeenth-century German mathematician who formulated the laws of planetary motion. "Kepler was a mystic and astrologer. But in this big mass of nonsense there are the three Kepler laws."[36]

Similarly, Zwicky advanced many ideas that were considered outré by his colleagues. But the things he was right about towered over his flights of fancy. The expanding universe just happened to be one of the things he was wrong about.

Many years later, after Zwicky was gone, a Caltech lecturer captured this character trait in a speech to visiting alumni. Of all the great minds that had studied and worked at Caltech over the decades, literature professor J. Kent Clark singled out Fritz Zwicky to demonstrate the quality of pure, unbridled inspiration that was so much a key to discovery. Zwicky, he said, was as much "a natural wonder of Switzerland as Mont Blanc."[37]

"Unlike you and me, he was not afflicted by modesty, he frankly told you he was a genius," Clark said. That was only partly true. In private, Zwicky could be modest. But he saw little reason to be shy about his ideas and accomplishments.

"Unlike you and me, he was a genius," Clark said.

Referring to Hubble's expanding universe, Clark recalled that Zwicky was known to call those who believed in it "horses' asses." This was thoroughly in character. Unlike the cautious language used in the scientific world, Zwicky's verbiage was uniquely colorful. He had many names for his multitude of enemies, from yes-men, windbags, inconsiderate careerists and even "toadeaters."[38] One of his most colorful epithets was "spherical bastard," a term that employed the language of physics to describe a person that, no matter what frame of reference was used, remained a bastard.

"To this day I can't take the expanding universe theory seriously," Clark said. "How can you accept a theory that is embraced by horses' asses?"

When it came to the expanding universe, Zwicky revealed his greatest weakness, that of failing to take stock when he was wrong. It might have been partly due to a resentment of Hubble's celebrity. It may also have been due to the fact that Zwicky had yet to make his own mark on the scientific world in any meaningful way. Mostly, it was because he was Fritz Zwicky.

Ironically, the inspirational genius Clark praised showed up in another section of the very same paper for *Helvetica Physica Acta* that proposed the tired light theory. In what seemed to be throwaway remarks about the so-called Coma cluster of galaxies, he made a truly remarkable assertion that would resound down the years.

The Coma cluster, Zwicky said, contained 800 galaxies (today the figure is known to be more like a thousand) and is about 2 million light-years across. He calculated the total mass of the system, all the millions of stars and interstellar gas and the rest, at 1.6×10^{45} grams—a lot of mass, even when measured in grams. He then noted that the velocities of the individual galaxies in the cluster varied by as much as 2,000 kilometers per second. This wide variation meant the cluster should not have stayed together, at least according to physics as it was understood then, and now.

To keep the entire cluster from flying apart, the "average density in the Coma system would have to be at least 400 times larger than that derived on the grounds of observations of luminous matter," Zwicky wrote. "If this would be confirmed we would get the surprising result that dark matter is present in much greater amount than luminous matter."[39]

No one had heard of dark matter,[40] and Zwicky didn't define it. What he meant was that there had to be a lot of hidden stuff in the cluster to have enough gravity to keep the whole caravan streaming along in formation.

This finding, he wrote, "harbors a problem that is not yet understood."[41]

Eight decades later, it still isn't. Around the world, in space and deep underground, scientists are working hard to find the answer to the problem of dark matter. Unraveling the mystery is today a leading goal of physicists worldwide.

The problem is, they don't know what they are looking for. The only thing scientists agree on is that, whatever it is, the dark matter particle— if it is a particle—is about as shy as it's possible to be. The kid too nervous to ask a girl to dance is a wild extrovert compared to dark matter. It won't have anything to do with anything we know, except gravity. Even Zwicky, whose 1933 calculation convinced him it existed, couldn't say where to look for it.

"Fritz proposed dark matter but he had no idea what it was," said Sunil Golwala.[42] A professor of astrophysics at Caltech, Golwala is a stocky man with wire-rimmed glasses, who wears his facial hair in the scruffy, modern style popular with Hollywood actors. He also happens to be an expert on dark matter.

Many scientists believe that dark matter was created in the first fraction of a second after the universe came into being, around 13.8 billion years ago. As the cosmos cooled and began to settle into clumps of stars and galaxies and clusters of galaxies, dark matter was there in the background, providing the scaffolding on which the superstructure was built. While it's invisible to the electronic eyes and ears that humans have trained on the universe, it is more responsible than any other force or object for the way the night sky looks to us.

"Without dark matter, it is possible we wouldn't even be here," Golwala said. Today, it's known that dark matter—Zwicky's unknown but essential element of all that exists—amounts to 24 *percent* of the universe. The stars and planets and intergalactic gas and the rest of the flotsam we can see and measure makes up only 4 percent. The rest is dark energy.

Although Zwicky proposed its existence, almost no one believed him. He was already known to be too bold, too rash with his surmises. He had been gone four years when the scientific world awoke to undeniable evidence that dark matter exists. Even then, many scientists refused to believe it. It's easy to blame scientists of the past for their failure to appreciate Zwicky's insight for nearly a half-century. But he didn't think enough of it himself to spend much time turning it over in his mind. The entire matter was disposed of in one small passage of his redshift paper. He was at the time pursuing some other ideas, which would bring him almost as much fame, and bother, as Einstein's theories brought him.

Zwicky saw the results of that notoriety when Einstein himself came to visit Caltech in the early 1930s, occupying a small apartment at the Caltech faculty club, the Athenaeum. He visited Mount Wilson and struck up a friendship with Zwicky that led him to endorse his impetuous colleague. His most frequently cited remark was that Zwicky was "one of the most promising young physicists," a phrase that appeared many times in print when reporters attempted to show that one of Zwicky's ideas wasn't as far-fetched as it might seem.

Fritz Zwicky was delighted by the praise, but his attitude toward Einstein remained a fraught one. On the one hand, he considered Einstein the one exemplary scientist of his era, the one man of genius he looked up to and whose theories he did not openly challenge. On the other hand, he couldn't help trying to find a weakness in the man. More than once, he would say that the worst thing Einstein could have done was to take the lifetime position at Princeton's Institute for Advanced Study the year after he left Caltech. At Princeton, Einstein had no teaching responsibilities, which was just fine with him, because he hated teaching. He spent much of the rest of his life working on the so-called theory of everything, ending with nothing.

The worst thing for a scientist, Zwicky argued, was to have nothing to do but think. For an energetic man like Zwicky, this philosophy made sense. Since he simply couldn't sit idly and reflect for long periods of time, having no responsibilities would have been worse than a jail sentence.

At the time of his yearly visits between 1931 and 1933, Einstein was at the height of his popularity as a public figure. Anticipation was at such a fever pitch in advance of his arrival in southern California that reporters consulted Zwicky about how to behave in his presence, learning "a number of interesting facts about the little Jewish genius who will soon be our guest."[43] Including that he would have little interest in listening to the ideas of ambitious American theorists. At Caltech, professors promised to keep him from "being pestered."

These good intentions went out the window when Einstein's ship docked in San Diego, on December 31, 1930. Children in blue and white sailor suits serenaded him, and thrust flowers into his hands, as two bands struck up lively tunes.[44] "The radios, the banquet tables, and the weeklies will never be the same," said the humorist, Will Rogers. "He ate with everybody, talked with everybody, posed for everybody that had any film left, attended every luncheon, every dinner, every movie opening, every marriage and two-thirds of the divorces."

Zwicky conveyed the same idea more succinctly, saying, "The journalists were after him as if he were the Holy Ghost or the devil incarnate."

When Einstein finally arrived at Caltech, he was "totally exhausted and disheveled. . . . But what annoyed him most was that almost every day he was invited by millionaires, politicians, scientists and colleagues in physics. He found these occasions either boring in the extreme or he simply ate and drank too much."[45]

Einstein shared some of his more morose thoughts with Zwicky, with whom he could speak openly, in German. "All those pompous people who thought they had achieved something major in science would have been horrified if they had known what Einstein actually thought of them," Zwicky wrote.[46]

The staff of the Norman Bridge Laboratory of Physics at Caltech, plus one distinguished visitor. Zwicky is seated, and crowded, in the third position in the front row. Two seats to his right is Robert A. Millikan, head of the laboratory, and next to him is Albert Einstein. This was Einstein's first visit to Caltech in 1931, during which Zwicky discussed his theory of crystal structure with the great man.

According to Zwicky, it wasn't until Einstein was driven up the coast, to be fawned over at one of the grand estates in Santa Barbara—even then it was a resort town populated by old-money Californios, installed in Spanish hacienda-styled ranchos overlooking the Pacific Ocean—that the professor made a pleasing discovery.[47]

"When he was back in Pasadena he couldn't wait to tell me," Zwicky said. "'Zwicky, I have found it.' I expected him to tell me that he had worked out an experiment, a cornerstone for a general field theory, combining gravity and electromagnetism, an old problem that he had been working on for 14 years, a problem that the greatest scientists had been tussling with for a whole century. But no! It was something different. What he had found, believe it or not, were two Mexicans, a man and a woman, both over 80 years old, who grew cactus plants. They had long since given up trying to cope with people or animals. . . . Their attempts had always gone wrong in some way or other. . . . So they had finally resigned themselves to just caring for prickly plants."[48]

The thirst for the simple life afflicts many people, especially those who have found the notoriety they once wanted so badly but then experience as a burden. There would be times in Zwicky's life when he wished for his own prickly plants. But just then his mind was fixed on a grander problem, the solution of which would bring him lasting fame. The physicist Zwicky, the crystals man, Millikan's hair shirt and Hubble's antagonist, was about to become Supernova Zwicky.

ON THE TRAIL OF ZWICKY'S GHOST

The Skeleton of the Universe

The problem with finding dark matter is that scientists, even now, don't really know what they are looking for. Zwicky himself didn't know how important his prediction would become. In his last years, he rarely even brought up his old suggestion that there was a lot of missing mass in galaxy clusters. While he lived long enough to see his neutron star prediction verified, he had been gone four years when Vera Rubin offered convincing proof that dark matter existed.[1]

Rubin, like many future astronomers, was fascinated early by the stars, studying the heavens at the age of ten outside her bedroom window in Washington, DC. Her father was skeptical, because girls were not supposed to be interested in the hard sciences in the 1930s. But he helped her build her own telescope and chaperoned her to meetings of amateur astronomers. When she graduated from Vassar in 1948, which was a women-only institution at the time, Rubin was the only astronomy major.

After earning her PhD, Rubin passed on a safe teaching career and began studying the dynamics of galaxies. As Zwicky revealed to the world, galaxies come in many shapes and sizes, from a few thousand stars to hundreds of trillions spread over mind-bending distances, sometimes more than 300,000 light-years across. Rubin was particularly interested in spiral galaxies, the giant, rotating whirlpools in space, like our Milky Way.

Just a handful of years after Zwicky's death, she was studying stars on the edge of one spiral galaxy when she noticed an amazing thing: stars on the outer edge were orbiting just as fast as those nearer the center, which was impossible unless there was a lot of invisible material on the boundary of the galaxy. That extra mass was necessary to allow the outer stars to keep pace with the inner portion of the galaxy. At first, she couldn't believe her discovery. But after checking one spiral after another, she finally accepted her own results. There must be a tremendous amount of unseen matter in the halo of the galaxy, she said in 1978, as much as ten times more than the visible stuff.

"What you see in a spiral galaxy is not what you get," she famously said.

At first, other scientists refused to accept her findings, just as they did when Zwicky proposed the existence of dark matter. But the evidence kept accumulating until, finally, most of the scientific establishment fell in line and accepted the fact that dark matter is real.[2]

Dark matter existed. Then the question became: What was it? Finding the answer to that riddle was what a sprawling lab complex at the bottom of the old Soudan iron mine in Minnesota was designed to do. Getting down there required a rumbling, shaky, 2.5-minute descent in the same conical cages that the miners took a century ago to get at the iron ore. Stepping out on Level 27, a half-mile underground, one was confronted by a football field-size cavern with gunnite-covered walls and ceilings vaulting forty-three feet above the floor. At the far end was something that looked like a giant steel archery target. That's actually not a bad way to think about the MINOS device. Consisting of 486 steel plates (they planned for 500 but ran out of money) stacked on end like a shelf of compact discs, each a half-inch thick and 28 feet across, the three-story-tall experiment was designed to capture and examine one of the tiniest bits of nature.

Every two seconds, the Department of Energy's Fermilab, 455 miles away, shot 2 trillion neutrinos at the 5,500-ton bullseye in Soudan. The beam started out as thin as a human hair, but by the time it reached northern Minnesota, it was a miles-wide gush of particles. "We're shooting a lot of neutrinos into Canada," said Marvin Marshak, a scientist with a sense of humor as dry as his state is wet. "Maybe we're shooting some moose."[3]

The moose would never know. The chance of a single neutrino colliding with the nucleus of an atom in the 109-foot-thick stack of steel plates—forget about a moose—is about the same as rolling a marble through the solar system and hoping to hit a planet. Over the years, MINOS—which stood for Main Injector Neutrino Oscillation Search—averaged only a handful of collisions per week.

Marshak, who taught physics at the University of Minnesota, was the man most responsible for this huge underground complex. "The person really responsible is my wife," Marshak corrected, in his round Midwestern drawl. "I was out scouting mines in Colorado and when I got back she said—her name is Anita Kolman—'I'm six months pregnant and we have a 3-year-old. What are you doing running around the country? Why don't you look at mines in Minnesota?'"

So he did. Getting the cooperation of the state for permission to blast out the cavern in Soudan wasn't the biggest challenge he had to overcome. "At the time, the federal government was looking for places to store radioactive waste, which they're still doing, I guess. There was some suspicion that that's what we were up to. I remember there was a meeting at a local school in 1979. About 200 people showed up, which was about a quarter of the population."

To allay suspicions, while he described his project, and why it had to be done deep underground, he held his infant son in his arms. "I never told him he was used as a political prop," Marshak said.

In the strange world of subatomic particles, neutrinos have always been one of the oddest members of the family. Shy shape-shifters, they are so elusive that the man who theorized their existence, Wolfgang Pauli, bet a case of fine French champagne that no one would ever find what he called Particle X. Pauli is most famous for the Pauli Exclusion Principle, which says two electrons can't be in the same area, doing the same spinning dance steps, at the same time. The principle was so important in understanding the structure and behavior of atoms that Pauli was given the Nobel Prize.

He was, however, not much of a gambler. He had to pay off his bet when the missing particle was found. Enrico Fermi, popularly known for creating the first human-made nuclear chain reaction in Chicago Pile-1, under the football stands at Stagg Field in Chicago, in December 1942, was given the right to name the new member of the subatomic family. Fermi chose the Italian term for small, neutral particle: neutrino.[4]

For a long time, it was believed that neutrinos had no mass. But as scientists learned more and more about them, they began seeing strange behaviors. Some physicists suggested neutrinos might oscillate as they travel, turning from one type into another as they flit around their tiny world.

What makes that so important, and not just for particle physicists trying to construct a taxonomy of the subatomic world, is that if neutrinos change flavors, they must have mass. Some astrophysicists immediately realized that, if so, they might solve the puzzle that Fritz Zwicky had posed decades earlier, when he said there was a lot of missing mass in the universe. After all, there are an awful lot of neutrinos in the universe. Maybe they were Zwicky's dark matter, the stuff that held galaxies together and was more than any other object responsible for the way the universe looks to us.

The oscillations MINOS was looking for were too small to add up to the enormous numbers needed for dark matter. But the work proved that the old Soudan cavern was a good place for such delicate experimentation. Which is how the CDMS detector—for Cryogenic Dark Matter Search—ended up at the bottom of a hole in far northern Minnesota.

* 6 *

NEW ALLIANCES, NEW PHYSICS

FRANKLIN DELANO ROOSEVELT'S election in 1932 gave hope to a nation beleaguered by the Depression and the moralizing of Prohibition, which, for a dozen years, had prevented the miserable and unemployed from escaping their troubles in strong drink. There remained a lot of trouble ahead. The year 1933 would be the very bottom of the economic trough, when unemployment hit a remarkable 25 percent. The country's gross domestic product would shrink to half what it was only four years earlier. The Great Migration west was fully under way, and poverty would not relax its bite for some time.

But with Roosevelt's election, people dared to believe that change was on the way. "After the election of Roosevelt and the prospect of unlimited quantities of schnapps, life here has taken on a fresh impetus," Zwicky wrote to his Swiss friend, "and everyone is looking forward to an inebriated future."[1]

The jaunty tone of Zwicky's note to Rösli Streiff reflected not only his relief that he and his friends would no longer be breaking the law when they tippled after hours, both on and off campus. The man who had begun referring to himself as a "lone wolf" had recently entered two important partnerships that would change his life and alter the course of scientific history.

The first began at a Caltech event in October 1931, where he met a 27-year-old socialite named Dorothy Gates. They hit it off immediately.

Dorothy represented everything with which a young foreigner trying to make his mark in America could wish to adorn his life. She was attractive, if serious, with a handsome face, thin lips, and a steady gaze that conveyed depth and determination. Despite his family's long history and his own father's accomplishments, Zwicky was conscious of the fact that he remained an outsider in America. Dorothy offered entrance to social circles to which a simple professor, even one who had already built a reputation among his peers, would not normally be admitted.

Hers was one of Pasadena's, and southern California's, most respected families. Her grandfather was one of California's educational pioneers. Her father, Egbert,[2] gave up a career in medicine to get into the ranching and mining business in Mexico and Arizona. The gold strikes of a couple of generations earlier were still fresh enough in memory for entrepreneurial people like Egbert Gates to keep trying their hands. Fritz Zwicky did the same. Shortly after meeting Dorothy, he invested some of his hard-earned teaching dollars in a gold mine in Mexico. It was a costly education for the young professor, who lost several hundred dollars before the enterprise was abandoned for lack of gold.[3]

The new gold mine in the early 1930s was California real estate, and Egbert Gates was naturally drawn to it. He became president of the Huntington Beach Company, founded by railroad pioneer Henry Huntington. When, in 1920, rich petroleum reserves were discovered, the company got to work drilling—quite successfully. Everyone connected with the firm became rich.

Gates's newfound oil wealth enabled him to try his hand at politics, winning election to the upper house of the state legislature. A rock-ribbed Republican, his values were deeply rooted in protestant economy and rigorous morality. His most notable achievement in the state senate was to spearhead a campaign in 1923 to change the law so that criminal trial juries could convict with only a three-fourths majority. The measure was inspired by one of those periodic paroxysms of panic over lawlessness that juries didn't seem able to stop. During the Gold Rush, Los Angeles suffered through the highest murder rate in the country. Crime was so bad that some eastern churches decided to forgo tribal Africa, instead sending soul-saving missionaries to southern California. Prohibition had not improved the behavior of crooks, or juries, for that matter, in Gates's mind.

"In criminal practice matters are reaching a point where it is almost impossible to get a verdict," Gates told reporters.[4]

Little remembered, this was one of the most audacious assaults on the rule of law in the history of the state. Many judges signed on, urging that the matter be submitted to the voters. The measure died only when the lower house of the legislature declined to place it on the ballot.

Egbert Gates faded into history, along with his failed morality campaign, after suffering a fatal heart attack while visiting New York. His estate was valued at $470,000, which yielded sizable trust funds allowing Dorothy and her younger sister, Tirzah, to travel when they wanted in whatever style appealed.

Zwicky had been ambivalent about marrying an American. And about the estate of marriage in general, being, as he was, a modern-thinking man who cherished his freedom to go off climbing mountains and hiking deserts whenever the impulse struck. He still talked about returning home, even though the idea was growing ever more unlikely as he began delving into research that could only be done at the controls of California's great telescopes.

On the other side was the fact that, whether he admitted it or not, he had a deep desire to make a family. The loss of his own family had forced on him an independence that he treasured. But there remained a barely acknowledged desire to build around him a society that would replace the sense of peace and security he lost when his parents sent him away.

For Dorothy, the attentions of a man whose name and ideas were being chronicled in the popular press as well as scientific journals was naturally appealing. But she could be diffident. In fact, she was so reserved that some of Fritz's friends considered her too self-contained, even haughty, adapted as they were to more deferential women. Worse, her Swiss German was nonexistent.

There were deeper problems that would only grow as time passed. For one thing, Dorothy was a Christian Scientist, a religion that tends to blame physical ailments on spiritual infirmity. Zwicky, not particularly religious and downright hostile to the fundamentalist doctrine he found in America, could not understand that attitude. He had a deep respect for medical professionals and rarely failed to submit himself to their interventions, whether it was changing his diet after suffering a heart attack in his fifties while charging up a mountainside, or laying aside his pipe after smoking's association with cancer was discovered.

More significant was the fact that Dorothy didn't want children. In his early thirties, that didn't seem like a problem. But as time went on, and his own family in Switzerland died away, it would grow in importance.

Whatever his doubts, in the beginning, what Dorothy offered to a lonely man in a foreign country made up for any perceived shortcomings.

By March of 1932, the two were husband and wife, having eloped to Santa Cruz, where Tirzah's soon-to-be-husband, Nicholas Roosevelt, a cousin of former President Theodore Roosevelt, had taken up residence on a gorgeous stretch of Big Sur called Point of Whales. "Pasadena society and scientific circles were given a big surprise yesterday in the form of a simple little announcement from Mrs. Egbert J. Gates, a member of one of Pasadena's 'first families,'" is the way the local paper put it.[5]

Zwicky, the article noted incorrectly, had come to the United States in 1927. "Since then his researches have brought him world recognition." That was an overstatement, but within a short time the remark would be considered prescient.

Zwicky's friends and patrons at the Swiss Federal Institute of Technology were quick to respond to the news. "I am overjoyed that you have found the woman with whom you wish to spend the rest of your life," wrote Paul Scherrer, Fritz's former professor. "If I know you she must be a very special girl."[6]

The great mathematician Hermann Weyl was also pleased. "What a surprise and what a welcome surprise! Sincere congratulations from both of us on your, to quote Einstein, 'having taken the plunge,'" wrote Weyl's wife, Hella.

Zwicky had been especially fearful of Rösli Streiff's reaction. Although theirs had been a friendship maintained almost exclusively through letter writing and would never be much more than that, he wrote her what he called a difficult letter about Dorothy in May 1932.

"I married a girl from Pasadena, and I am embarrassed to say that our long-planned tour in Valais can't really go ahead. I hope you are not too annoyed that I didn't write to you earlier. I first met my new wife in October last year, and two months ago we decided to get married."[7]

At that time, even in forward-thinking social circles in Los Angeles, where people ignored the Prohibition laws and accepted divorce as a con-

MRS. FRITZ ZWICKY
—Ray Huffstudio

PASADENA GIRL
WEDS SCIENTIST

The April 7, 1932, announcement of the marriage between Dorothy Vernon Gates and Fritz Zwicky, "one of the most outstanding young scientists in the world." Zwicky was ambivalent about marrying an American, which meant severing one more connection to Europe, and was embarrassed to admit the fact to his Swiss admirer, Rösli Streiff.

sequence of the complicated life led by modern human beings, it was still not done for unmarried people to cohabitate. There was no choice but to marry. The hastiness of it—neither her family, except her sister, nor his were invited—produced some second thoughts.

"This decision caused both my wife and myself to give up a lot of things dear to us. It seems this cannot happen without some suffering," he wrote. "The worst irony for me is that I didn't meet you sooner, even though we both grew up in Glarus. I can only say that you have become dear to me through your letters."

These addendums were odd. Was he trying to keep Rösli on the hook, in case the marriage failed? His letter certainly displayed more ambivalence than delight. Although he was careful to include Dorothy in his comment about having to give up hobbies, it is more likely he was speaking mostly about himself and his freedom to go exploring and roughnecking in the woods with friends.

On their honeymoon to Europe, Dorothy gamely accompanied him on a trip to the 10,000-foot Gornergrat mountain, where he took a picture of her on a rock outcrop with snow in the background. Wearing a flapper hat, she was dressed more for a night out to Musso & Frank Grill in Hollywood than for a grueling hike in the mountains. Her smile was broad enough that it's not unreasonable to think she might even have enjoyed the outing.

She would further prove herself on a two-month trip to Alaska, on which the couple, accompanied by Dorothy's younger brother, Howard, a student at Stanford University, spent a lot of their time camping. Zwicky told the local reporters on his return that the salmon fishing was glorious and the rivers were fresh and clean. They encountered plenty of gold miners, the fever having been transferred north from the played-out California gold fields. But even he had to admit that the constant rain became "no fun" after a while.[8] Mrs. Zwicky was not solicited for her opinion of the expedition.

After his return, he sent Rösli a note apologizing for not writing. There had been an "immeasurable amount to do" at the Institute. This was partly because of all the traveling he and Dorothy had done after the wedding. Then they had to find a house. After searching fruitlessly through one Pasadena neighborhood after another with "such terrible architecture that your head stands lopsided after three days," they finally found a rental near Caltech.[9]

His duties at the institute had also expanded. In addition to lecturing in physics, as an associate professor, he had begun teaching astrophysics, the field in which Millikan had challenged him to make his name. The assignment was, as much as anything, a backhanded compliment. Zwicky was young, with no status. He would not become a full professor of astrophysics until 1942. That he would turn the new assignment into a perch from which he would survey the universe and bring back strange beasts likely never occurred to Millikan.

"I did already have some knowledge in this field, but it was never enough for lectures just like that," he wrote to his Swiss friend, "so I really always have to prepare myself enormously."[10] His "Analytical Mechanics" course, required for a PhD in physics, would be remembered by his weary students as especially tough.

The second partnership he entered would at last bring him the achievement and recognition he so deeply wanted. Walter Baade's genius was every bit the equal of Zwicky's. Not in its leaps of imagination and insight, maybe. There were few who could match Zwicky there. Baade's intellect was entirely different, expressed in a surgeon's peerless use of the new astronomical tools to probe and excise truths that stared others in the face but lay beyond their ability to understand. Like a number of Zwicky's associations, the relationship would eventually fracture, leaving both men the bitterest of enemies. But while it existed, his connection with the German astronomer was one of the most fruitful alliances in modern scientific history.

Baade was the oldest of four children of a teacher and was five years older than Zwicky. He was born with a hip defect that kept him out of World War I and contributed to his premature death at sixty-seven. With jug ears, the ever-present pipe dangling from the corner of his mouth like a fishing pole, a protruding, eruptive nose, and thin, pursed lips that warned away the uninvited, he was a figure straight out of central casting for a mid-century professor or police inspector.

Walter Baade and Zwicky formed an extraordinarily successful partnership in the 1930s, startling the scientific world with their predictions of neutron stars and supernovae. Their relationship would eventually fracture, but both men would go on to make important contributions on their own.

Unlike the Swiss physicist, Baade was a trained astronomer, having graduated with honors from the University of Göttingen, which was arguably even more famous for producing great scientists and mathematicians than ETH in Zurich. It was to Göttingen that Einstein went after leaving the Swiss institution.

As he neared completion of his PhD thesis, Baade won a prestigious appointment in 1919 to the Hamburg Observatory, site of the 40-inch Zeiss telescope.[11] It was the largest telescope in Germany, but it was a poor instrument compared to the telescopes on Mount Wilson, afflicted by its location near the grimy, ancient port city on the river Elbe. One of the Hamburg Observatory's least impressive accomplishments was to claim to have seen water-filled canals on Mars. This "finding" contributed to the excited speculations in both the scientific and popular press about the existence of an advanced civilization on the red planet.[12]

Baade was well aware of the shortcomings of the Hamburg Observatory. He was also knowledgeable about the work being done by Hubble and the other young Americans at Mount Wilson. He yearned to see the place for himself and hoped to win a fellowship that would bring him to the United States, which he did, from the same Rockefeller Foundation that brought Zwicky to America. The fellowship paid seventy-five dollars a month, fifty dollars less than Zwicky received. Arriving in 1926, he spent a year building a reputation as "one of the most promising young astronomers of the world," in the words of Mount Wilson's director, Walter Adams.[13]

When the year was up, Baade returned to Germany, and its cloudy, dirty skies. He couldn't help comparing his instrument to the magnificent telescope he had put his hands to in California. Just as troubling, conditions in Germany were deteriorating; the Depression that arose out of the stock market collapse in America had spread around the world. Millions in Germany were out of work. Inflation was rampant, and the Weimar government was flailing. Germans yearned for leadership, and Adolf Hitler volunteered his services. All in all, the outlook, figuratively and literally, was gloomy. So when Adams came calling with a job offer in 1931, paying $3,300 a year, Baade eagerly accepted.

The Pasadena offices of the Mount Wilson Observatory, which was under the direction of the Carnegie Institution, were only blocks from

Caltech. It was natural that Baade and Zwicky would meet. When they did, they became fast friends and soon, collaborators. Though their falling out later would be as explosive as the eruptions they chronicled, in the beginning the two were well suited to each other. Baade was the cautious observer who now had access to the world's greatest astronomical instrument, the 100-inch Hooker telescope. In contrast, Zwicky was the instinctive artist, capable of dazzling feats of insight and confident enough to stand against the opinions of the experts. Each needed the other. Baade's caution required Zwicky's boldness to assail conventional wisdom; Zwicky's rashness needed Baade's measured diplomacy and facility with big telescopes.

Although known for his good humor, Baade could be acerbic, even physically confrontational. He once rolled on the floor of a dark room with another astronomer, who was, yes, afraid of the dark; and he dismissed what he called "weak pets" in safe research positions. It was an attitude that in the beginning made him and Zwicky natural allies.

Crucially, both men were interested in the largest structures and most powerful energies in the universe. By late 1933, they were ready to unveil some of the most daring ideas in the history of astronomy.

Zwicky stole a march on the Stanford conference at which he and Baade unveiled their cosmic ray and neutron star theories by announcing his new ideas before the sessions even began. Days before the meetings, newspapers all over the country, from Pennsylvania to Oregon, from Kentucky to Wisconsin, heralded the new theories, and, not inconsequentially, the theorist.

"Cosmic Rays Produced by Blasted Stars," said the *Daily Capital Journal,* in Salem, Oregon. The paper misspelled super-nova (the term then used) as super-novaf. "The proof of this theory will have to wait until another super-novaf appears in the earth's star system, and that may be one year or a 1,000 years from now."[14]

Reporters on deadline rarely mentioned Baade's name, though Zwicky did make a point to mention him in at least some of his interviews. Zwicky's announcement to the world before the conference was a masterstroke in public relations that a Hollywood publicist might admire, and it likely contributed to the resistance the men encountered at Stanford.

In the aftermath of the Stanford conference, Zwicky and Baade produced several more papers, tidying up their conclusions and adding details to their theories about cosmic rays and supernovae. They also adopted a more cautious tone, no doubt as a result of the outrage their ideas had generated in scientific circles. In one paper, they even admitted being "perhaps too optimistic" in their estimates of the intensity of the cosmic rays produced by supernovae.

On the subject of neutron stars, they had also become more cautious, saying they had only "tentatively suggested that the super-nova process represents the transition of an ordinary star into a neutron star. If neutrons are produced on the surface of an ordinary star,[15] they will 'rain' down towards the center if we assume that the light pressure on neutrons is practically zero. This view explains the speed of the star's transformation into a neutron star. We are fully aware that our suggestion carries with it grave implications regarding the ordinary views about the constitution of stars and therefore will require further careful studies."[16]

This paper listed Baade's name first, and it likely reflected his own fears. Even though, over the ensuing decades, the main body of the scientific establishment refused to believe in neutron stars, Fritz Zwicky never yielded for a moment to the doubters.

This preternatural confidence also showed itself in a series of papers he wrote around this time—1933 and 1934—on an entirely different, and far more speculative, subject. In them, he challenged the very basis of the scientific method by proposing a new type of reasoning. This detour into philosophy revealed his unique problem-solving approach; it also lay at the heart of his incendiary conflicts with the guardians of mainstream science over the next four decades.

In this new type of reasoning, he argued for the "principle of flexibility of scientific truth,"[17] saying science and scientists must not rely solely on measurement and reproducibility. These principles were, of course, bedrock values of science, both then and now. Science holds that the only way to know something is true is if you can reproduce the same results from the same experiments, time after time. To Zwicky, this was unpersuasive. Truth, he said, was a lot more spongy and slippery than that. It's not clear how he arrived at this idea, but it seems he took inspiration from some of the confounding elements of quantum mechanics. The Heisenberg Uncertainty Principle, which held that one could not measure both the position and momentum of an elementary particle, had been introduced only six years earlier, and the scientific world was still wrestling with its implications.

Years later, Zwicky expanded on this line of attack by arguing that even the so-called physical constants could not be relied upon. Since measurements like the mass of the electron depended on the distribution of matter and radiation in the universe, and since that distribution must change over a billion years, the mass of the electron must change with it, he argued.

Zwicky went still further, arguing that the "so-called laws of logic are therefore themselves subject to the flexibility of scientific truth." Even worse, to mainstream scientists, Zwicky said science should not be quick to cast aside even metaphysics, which might hold truths that science could not probe.

All these ideas were essentially an attempt by Zwicky to look inside his own mind and to find out how it worked. In time, the process would lead to his discovery of what he would call his Philosopher's Stone: a system of logic and truth-finding he called Morphology. This wasn't New Age spiritualism, but it did carry a faint whiff of flower child utopianism, with its rejection of traditional forms of logic. Predictably, it set off a heated controversy, in particular with the Yale physicist Henry Margenau. "Zwicky suggests that . . . one must question every categorical statement," Margenau wrote. Quoting a Latin proverb, he sneered that, "to doubt everything or to believe everything are two equally convenient solutions; both of which dispense with the necessity of reflection."[18]

Zwicky responded to Margenau's challenges in character, saying he would debate "any such proposition which Mr. Margenau chooses to advance."

Zwicky was clearly feeling his oats. And when he did, he was given to grander pronouncements than usual. When the *Los Angeles Times* asked him to help them predict what life would be like in fifty years—Zwicky's foresight, the *Times* said, "is clearer than most persons' hindsight"—he spared no extravagance, saying that scientists, theologians, free-thinkers, dreamers, and even "nuts" would work together to approach the unknown in a friendly fashion.[19] Several of his other predictions—that new optics would allow people to see through fog and space and that ways would be found to store the sun's energy—were prescient. His hope that nonscientific efforts to find nature's truth would be honored went unrealized. Nuts are still nuts.

One fact about exploding stars that Zwicky fretted over was that neither he nor Baade had yet observed a supernova in the act of immolating itself. All their theorizing had been done on the basis of old records and mathematical suppositions. Zwicky immediately set about remedying the problem. Because he was not a member of the staff at Mount Wilson at the time, he decided to make his own observatory—on the roof of Caltech's Robinson building, where he had his office.

Only erected the year before the Stanford meeting, the Robinson building was topped by a scale model of the dome for the 200-inch telescope that would be constructed on Palomar Mountain. It was not as if Zwicky was climbing to the roof of his house and setting up a tripod. Still, his 12-inch telescope, with a 3.5-inch Wollensak camera attached, was not much larger than backyard telescopes today—a poor instrument with which to conduct a broad survey of the sky. He admitted many years later that his rooftop venture was "accompanied by the hilarious laughter of most professional astronomers and colleagues at Caltech."[20] He made up for the lack of good tools with his usual determination. For months at a time, from 1934 to 1936, he canvassed the sky, looking for an erupting supernova. Meanwhile, Dorothy sat at home, alone, an astronomy widow. This was not an unknown species. It was accepted among sky watchers that their profession could be hazardous to marital contentment.

One possibly apocryphal story quoted a frustrated wife complaining, "I would have divorced you, but you were never home long enough to discuss it." Dorothy, with her wide circle of friends and the resources available to travel abroad whenever the impulse struck, put up with it longer than many did.

Zwicky braved the stress on his marriage, not to mention the discomfort of sitting nightly vigil on the roof of a classroom building in Pasadena. But the universe refused to cooperate. His searches were all in vain. His instrument was simply unsuited to the task. The first to find an erupting supernova following publication of Zwicky's, and Baade's, theories was Hubble. He found it in January 1936, in the Virgo cluster of galaxies, just where Zwicky had predicted it would be found, two years earlier. This was a densely populated cluster 30 million light-years from Earth.[21]

The supernova was too faint to be seen with the naked eye, and Zwicky would never have found it with his poor instrument. But this was the first notable success for his speculations. Now the wait began to find the next one. There was no way to know whether it would be months, years, or decades before the next exploding star appeared.

Later that year, still frustrated in his searches, Zwicky and Dorothy traveled to Europe to attend a meeting of the International Astronomical Union, in Paris. While there, they visited Fritz's family in Switzerland. The dutiful Pasadena journalists were happy to report on the trip, relaying word that the traffic in Europe had become as congested as it was getting in Los Angeles. The price of fuel was worse. At a time when you could buy a gallon of gas in California for ten cents—and rent an apartment for twenty-four dollars a month—they paid eighty cents a gallon in Italy.

More important than the state of auto transportation was his fateful encounter with Bernhard Schmidt. In a field filled with strange birds, Schmidt might have been the strangest of all, particularly in his choice of plumage. He worked in a formal cutaway coat and striped pants, as if he were about to attend a formal dinner. Which, being painfully shy, he never would have done.

As a child, Schmidt lost an arm experimenting with explosives.[22] So ended an interest in munitions. Astronomy took its place. As an adult, Schmidt was often seen pacing his Hamburg suburb along the Elbe with a cigar projecting from his mouth at an angle so steep it looked as though he would set his bowler hat on fire. Money meant nothing to him. He turned down many high-paying jobs to pursue his goal to build the perfect astronomical instrument.

In the early decades of the last century, conventional telescopes suffered from a serious problem. The parabolic mirror, which concentrated starlight to a focal point where a photograph could be taken, produced fuzzy images on the edges. Even the great 100-inch Hooker telescope had this defect. Taking photographs with this kind of telescope required great patience. Obtaining good plates could take up to fifty hours, spread over ten successive nights.

Schmidt, during one of his ceaseless walks around Bergedorf, had a sudden inspiration: what if one installed a correcting plate that would bend light entering the telescope in just such a way that the image reflected back to the eye and the camera would be perfectly crisp and clear? There would be several different correcting plates for different telescopes. Just as important, the correction would allow a wider field of stars to be photographed, meaning great portions of the sky could be studied at once, making the Schmidt model a perfect scouting tool.

He first described his idea to none other than Walter Baade in 1929. The two men were aboard ship in the Indian Ocean, on their way to view an eclipse. At the time, scientists were still testing Einstein's theory that gravity bends light. An eclipse presented an excellent opportunity to measure the light rays from far-off stars as they passed the sun, without the blinding interference of Earth's parent star. Baade was then working at the Hamburg Observatory, biding his time until he could catch on at Mount Wilson. During the ocean voyage to the Philippines, Schmidt boasted that he had designed "the perfect mirror."

"You will have to build it when you get back," Baade said.[23]

Schmidt did build it. Then he offered it to an astronomical world badly in need of just such an instrument. Only nobody wanted it. At a time when bigger was better, when the 100-inch Hooker dominated astronomy, his modest, 14-inch telescope seemed puny. One of Zwicky's greatest regrets would be that he didn't buy the telescope when Schmidt offered it for the relatively modest sum of $1,100.[24]

After meeting Schmidt during his auto tour, Zwicky returned to Pasadena with a passion to build a telescope on Schmidt's design. Lacking the money, he turned to Hale, who was not enthusiastic. Though he certainly appreciated his friend's energy and industry, so like his own, Hale was not much interested in Zwicky's researches into exploding stars. To Hale, who didn't realize how fundamental such events are to understanding the great forces in the universe, focusing on the violent deaths of stars seemed prurient. In his opinion, people who looked for supernovae were like California gold miners, trying to strike it rich. "They are not very respectable," Hale said, according to Zwicky.

"So I almost gave up," Zwicky said later, "and I said: 'Dr. Hale, it all depends on who searches.'"

"Okay," Hale laughed. "Twenty-five thousand dollars."[25]

It took a year to make the new telescope, which Zwicky pitched in to help build.[26] The first instrument to go up on Palomar Mountain, it saw

Zwicky at the eyepiece of Palomar Mountain's 18-inch Schmidt telescope, which he helped build. The picture was taken in the summer of 1937, as Zwicky was "beating the tar" out of the sky searching for supernovae. In March, he found his first, which stimulated worldwide interest in exploding stars.

first light in September 1936. Its correcting lens was only eighteen inches in diameter; not much, perhaps, when compared with the big Hooker. But the Schmidt could survey a starfield nearly nine times as large as other reflecting telescopes.

This made it perfect for the mapmaker's task of surveying great expanses of space in one go. In its working life, the 18-inch-Schmidt and its larger cousin, which had a 48-inch correcting lens, would discover dozens of supernovae, chart stellar populations, and track many asteroids and other minor bodies in the solar system.[27] It would help prove that galaxy clusters, the tendency of giant star systems to gather in great herds, was the rule rather than the exception in deep space. The Schmidt was key to developing the first detailed map of the great starry sea surrounding our Earth's small island.

When the telescope was at last ready, Zwicky began to spend three days a week there, inching his way along a rutted road high into the forest. An intrepid reporter described the conditions when he hazarded the trip in August 1936. The hundred-mile distance cost him 302 miles, thirty of them over treacherous dirt mountain roads, thronged with cattle.[28] When the road washed out, which was not uncommon, Zwicky had to be towed down the mountain by tractor. When she would go, he took Dorothy up with him; the only semi-permanent residents at the observatory site, the couple stayed in a small white-painted stucco cabin. Years later, remarried and a father, he bought another cabin and made it as comfortable as he could, with a heavy cast-iron stove that the family huddled around when the mountain was buried in snow. But at this early point, the conditions at Palomar were reminiscent of his mountaineering adventures.

For recreation, he built a ski ramp—piling up snow against the 18-foot dome—and began trying it out. His best leap was fifty feet. Not bad for an astrophysicist who was nearing 40 years of age.

This was the beginning of the period when Zwicky, as some described it later on, "began beating the tar out of the sky." Night after night, for weeks at a time, he took photographs, focusing on 132 separate areas of the sky. Then he would pore over his photographic films like a jeweler, searching for the sudden flare-up of a "temporary star," or, as he put it after Hiroshima, "atom bomb stars."[29]

In the winter, it was so cold that his fingers grew stiff and he had trouble manipulating the controls. Telescope domes could not be heated, because any change in temperature changes the optics of the mirror, as well as setting up air currents inside the dome, which spoil the seeing, as astronomers call it. The thick cloth coat he wore on his climbs could not protect his hands. He couldn't use gloves and work the controls. All he could do was suffer through it.

"In spite of all I have already made about 300 fine photographs with the Schmidt telescope, and processing this material will keep me busy for many months," he wrote to Rösli.[30]

He bagged his first supernova in March 1937, in the galaxy NGC 4157. Then in late August his research made a splash when he found the brightest supernova of the century.[31] In a breathless telegram to Harlow Shapley, now ensconced at the Harvard College Observatory, Zwicky said, "Bright Super Nova Magnitude 8.5 In Index Catalogue 4182 Observed Palomar August 28 (STOP) Invisible May 31 But Appears On Several Later Plates (STOP)."

Back at Mount Wilson, Milton Humason, Hubble's assistant, who was by now working with Zwicky, confirmed the discovery with the Hooker. The flaring star was located in a small system, 3 million light-years away. The explosion might be millions of years old, but it was breaking news on Earth.

"The discovery results from a systematic plan for the detection of the outburst of these amazing objects," went a press release reflecting the worldwide excitement over exploding stars, "which at their maximum brightness far surpass the luminosity of any other type of star."

"Astronomer's Hobby Wins Vindication in Discovery," is the way one newspaper characterized his achievement. "When the Schmidt was erected," according to the story, "Dr. Zwicky launched an extensive novae discovery program for the key to the riddle of cosmic rays, and the constitution and evolution of stars, and the discovery marks an amazing first step forward in the program."

Soon after, the discoveries were coming regularly.[32] His findings showed that his original idea, that a supernova occurs only once in a thousand years in a particular galaxy, was far too conservative.[33]

As he worked, he gradually improved on the techniques used to scan the night sky. One innovation he adopted was to coat his photographic plates in two different color emulsions. After superimposing them on one another, any change seemed to leap off the glass. This technique was widely copied. It wasn't the sort of thing that captured the attention of the public, but the discoveries he, and others, made using it did.

Another innovation was far more important to the field of cosmology. It concerned a little-noticed consequence of Einstein's theories. It was not Hubble or Zwicky or Baade who stumbled on the original idea, but a layperson identified in a press report as a dishwasher with an imagination. R. W. Mandl was not a dishwasher; in fact, he was a trained engineer, but his imagination was indeed impressive.[34]

Mandl understood enough of Einstein's theory of gravitation, particularly its ability to bend light, to wonder whether the effect could be used as a natural lens to magnify very distant objects. Back in 1912,

Einstein had calculated that if two stars were in exact alignment, the rear one would appear as a ring around the one in the foreground. Gravity bends the light around the nearer mass at the expense of other directions, thereby increasing the brightness of the distant source. He didn't publish the results at the time, because he didn't think they would be important.

Mandl first approached Vladimir K. Zworykin, then of RCA Labs, who was intrigued enough to suggest he talk to Einstein himself. In those simpler times, it was much easier to obtain introductions to famous people. So when Mandl showed up at the Institute for Advanced Study, where Einstein was busy meditating on the theory of everything, the great man agreed to listen. To Einstein, Mandl seemed a bit of a crackpot. One of his crazier ideas was that a gravitational-lensing event might have caused one of Earth's mass extinctions by spewing cosmic radiation on our planet.

"Your fantastic speculations associated with the phenomenon would only make you the laughing-stock of the reasonable astronomers," Einstein wrote to him.[35]

Eventually, Einstein relented and produced a paper on the effect. In a cover letter to his editor at *Science,* he referred to the paper as a "little publication, which Mr. Mandl squeezed out of me," adding that it was "of little value, but it makes the poor guy happy."[36] The paper Einstein wrote was titled "Lens-Like Action of a Star by the Deviation of Light in the Gravitational Field." Then he washed his hands of gravitational lensing, and Mr. Mandl, and went back to navel-gazing in Princeton.

That might have been the end of it but for Zworykin—who is today remembered for his pioneering work in television (he filed his first patent in 1923)—mentioning the incident to Fritz Zwicky in a letter. The two men did not know each other, but Zworykin was well aware by this time of the man behind exploding stars and other strange phenomena. Zwicky recognized something Einstein had missed: gravitational lensing would work as well with nebulae, or galaxies, as with stars. In fact, it would be a better tool. As Einstein realized back in 1912, the alignment of two stars happened so rarely that "there is no great chance of observing this phenomenon" in the real world. But with galaxies, enough was known about them in 1937 for Zwicky to realize that there was a far greater likelihood that two of them would align. Zwicky was sure this tool could help reach into the deepest regions of space, far beyond the range of even the largest telescopes.

In a short article of his own on the subject, Zwicky wrote, "I have made calculations which prove that extra-galactic nebulae offer a far better opportunity than stars to observe these gravitational effects." He said the technique could shed light on several problems then troubling astronomy: "(1) It would furnish an additional test for the general theory of relativity. (2) It would enable us to see at distances greater than those ordinarily reached by even the greatest telescopes. Any such *extension* of the known parts of the universe promises to throw very welcome light on a number of cosmological problems. (3) The problem of determining nebular masses at present has arrived at a stalemate."[37]

Here, he referred to his calculation that the observed mass of the Coma cluster was 400 times less than needed to keep it together. Proving that he had not abandoned his dark matter puzzle, he added, "This result has recently been verified by an investigation of the Virgo cluster."[38]

Zwicky never had a chance to witness the success of his idea. The first successful observation of lensing occurred in 1979, five years after his death. Since then, it has become one of the most important tools for observing objects and events reaching back almost to the beginning of time, just as Zwicky imagined.[39] The famous photographs of the Ultra Deep Field, taken by the Hubble Space Telescope, harken back to the time when galaxies were just beginning to form, only a billion years after the Big Bang. Our understanding of the benthic spacescapes of the universe owes everything to this idea proposed by Zwicky.

And as Zwicky hoped, the technique helped confirm the existence of dark matter, though that took a bit more time. Only in 2006, images taken of the so-called Bullet cluster of galaxies showed the effects of dark matter operating on a gigantic scale. The Bullet cluster, containing about forty galaxies, resulted from the smashup of two smaller clusters 150 million years ago, when allosaurs walked Earth and pterosaurs flew overhead. The collision occurred at a speed of 6 million miles per hour and produced temperatures in the clouds of gas accompanying the galaxies in the range of 100 million degrees.

The orbiting Chandra X-ray Observatory analyzed radiation from the hot gases and discovered that they were gathered into clumps around the clusters like lumps of cotton in a sagging mattress.[40] An invisible force was pushing the clumps around. It was not the first convincing evidence of dark matter, but it is one of the most compelling, since the massive clouds of clumped-up gas show up in images anyone can see and appreciate. Dark matter had at last stepped out of the shadows.

Since then, researchers have been able to chart dark matter on a larger, cosmological scale, discovering that the invisible stuff has been increasingly clumping up over the past 6.5 billion years. That's nearly half the age of the universe. It's to be expected in a universe that's flying apart while its individual galaxy members gather together for a sort of cosmological comfort against the yawning emptiness of the journey ahead. The discovery was a tremendous feat of modern cosmology.

Ironically, when he proposed using gravitational lensing to study galaxies, Zwicky hoped it would undermine the expanding universe. It didn't. But it did support one of his best and most insightful ideas. Had he known that, would he have made the trade?

Meanwhile, Zwicky's researches at Palomar were turning up more and more supernovae. Exploding stars seemed to be everywhere. Zwicky and Baade had said in 1933 that a supernova was rare, but by the end of the decade, Zwicky's total had reached twelve. Each discovery produced excited telegrams to Harlow Shapley, who announced them to the world.

"Supernova found by Zwicky at Palomar in Virgo cluster. . . . Observations urgent," shouted a January 20, 1939 telegram to Shapley from Walter S. Adams, the director of the Mount Wilson Observatory.

In 1937, Zwicky's discoveries were hailed as one of the greatest scientific achievements of the year. A Swiss journal crowned him "Supernova Zwicky."

The name was not intended to carry a double meaning. But in succeeding years, as his volatility began to color his reputation, other writers would use the same terminology. "On the theory that it takes one to find one, Dr. Fritz Zwicky . . . is splendidly qualified for his present post—planet Earth's head watchman for supernovae," went one profile, written in 1964. "For supernovae are highly explosive and controversial objects. And few scientists are more explosive and controversial than Professor Zwicky."[41]

This volatility contributed to the fracture of Zwicky's working relationship with Baade at the end of their decade of discovery. For the previous seven years, their relationship had brought them both fame and notoriety. With Zwicky at the controls of his 18-inch Schmidt telescope on Palomar Mountain, and Baade working at the 100-inch on Mount Wilson, they were among the most envied and celebrated astronomers of their day.

As late as August 1939, with a world war on the horizon, Baade was sending friendly tidings Zwicky's way from Paris, where he had gone to attend an astronomical congress. The congress "was very nice," Baade wrote, twelve people "around a table with an interpreter and two stenotypists." This shows the modest beginnings of cosmology. Later, meetings of the International Astronomical Union would draw thousands of attendees for conferences lasting two weeks.

One speaker gave a report on the energy of supernovae, which Baade said impressed only the "decomposed Russell."[42] This was Henry Norris Russell, whom Zwicky scorned as the "high pope" of American astronomy. He had written two popular astronomy textbooks,[43] but by 1939 he was in his sixties and considered old and irrelevant by young lions like Baade and Zwicky.

However, another participant presented a theory that Baade said was "not at all dumb." This was Subrahmanyan Chandrasekhar, for whom the orbiting Chandra space telescope was later named. One of ten children, born in 1891 in Lahore, Punjab, to a Tamil family, he was a nephew of the physicist and future Nobel Prize winner, C. V. Raman. Chandrasekhar had begun thinking about the life cycles of stars after graduating from Trinity College in England. At the Paris conference, he told Baade he had an idea that might explain why only some stars became supernovae.

The rarity of supernovae, Chandrasekhar said, according to Baade, "comes from this, that only stars in the mass range" of at least 1.43 times the sun "can make the transition to a neutron core. By the way, Chandrasekhar gave us full credit that our work for the first time mentions the fact that you cannot get around the formation of neutrons for supernovae. As already mentioned, you will get the details as soon as I get back."[44]

Chandrasekhar's idea was that only certain stars could gather enough mass, and gravity, to collapse to the colossally condensed neutron core.

This was later called the Chandrasekhar Limit, and contributed to his winning the Nobel Prize in physics in 1983.

Zwicky's and Baade's theories did encounter headwinds from one leading researcher at the conference, Georges Lemaître. Lemaître was a Belgian who carried on the once-common tradition in the Renaissance for men of religion to dabble in the sciences, in large part because they were often the only educated men around. In Lemaître's case, however, he began his career as a student of engineering, only later entering the seminary. It was he who first came up with the idea, though not the proof, of the expanding universe. At the Paris conference, Lemaître criticized Zwicky's and Baade's calculations about the energy released in a supernova. But, as it turned out, he had made a mathematical mistake. "We discussed this in depth in private afterwards, and I told him of your new work," Baade wrote. "Lemaître was very worried about having been unfair to us with his opinion."

In the end, the event came off very well, Baade said, with the reputations of the two scientists from California intact—enhanced, in fact. Even though it would be many years before the scientific community at large accepted the existence of neutron stars, some of the best minds of the day were being won over. Baade signed off with kind words for Fritz and Dorothy, saying he would be seeing them in a few weeks. These were some of the last cordialities exchanged between the two men.

Zwicky was resentful because he had not been invited to the conference in Paris, while Baade was a star there. Given that he was the one finding all the exploding stars, and that he was the face of the new world of high-energy astrophysics, his resentment is easy to understand. Just the year before, in 1938, Zwicky had come up with a new idea about what could happen inside a star even more collapsed than a standard issue neutron star, an idea that he thought would advance the understanding of the forces at work in the universe even more than his and Baade's earlier work.[45] As described in a note in the Caltech campus magazine, *Engineering & Science,* "Astronomers cannot prove it, but they strongly suggest that there are stars in the sky which, although shining, cannot be seen, according to a report by Dr. Fritz Zwicky."[46]

"Such stars would be of a type called collapsed neutron stars and would represent the lowest states of energy which matter could possess without turning into radiation," according to the report. This is likely what Zwicky would later call an "Object Hades," which he always claimed was a better term for a black hole.

Those who took Baade's side in the coming rupture cited the German's growing discomfort with Zwicky's brusqueness toward other scientists. Their methods of working were, in fact, greatly different. While Zwicky liked to work alone, or at least in charge of any team of which he was a member, Baade was a solid member in good standing in the broader astronomical community.

Baade's biographer, Donald Osterbrock, cites a supposedly telling dispute between Baade and Zwicky in 1936, in which Zwicky responded hotly to criticism of his work from Cecilia Payne-Gaposchkin. "Baade, who admired her work in general and never insulted anyone . . . soon began dissociating himself from Zwicky," Osterbrock wrote.[47]

The friendly 1939 letter from Baade casts doubt on that story. In fact, Zwicky's and Baade's friendship continued, and it might well have gone on longer than it did if it weren't for the arrival of World War II, which brought resentments flaring into the open and caused things to be said that could not be taken back. Zwicky's hatred of the Hitler regime was deep and often expressed. Baade, in contrast, was sympathetic to the German side, if not to Hitler personally. He and his wife, Johanna, nicknamed Muschi, were often heard using the term "we" when referring to the German war effort. In letters home, Baade made it clear he was not particularly fond of America, referring sarcastically to its "holy democracy." He railed against President Roosevelt's "cash and carry neutrality," the policy adopted to keep America out of the war in Europe, while selling arms to the combatants. Baade's dog menaced visitors unless they spoke German.

At the time, it was expected that foreigners working in American science would become citizens, an expectation Zwicky would later defy at great personal cost. Baade had no intention of becoming a citizen either. But when asked about it, rather than admit, as Zwicky would, that he was too attached to his homeland to sacrifice his citizenship, Baade gave the excuse that he had taken out papers but lost them.

There was enough suspicion about his sympathies that Baade was classified as an enemy alien after the outbreak of the war. This subjected him to curfew regulations, preventing him from being out on the streets of Pasadena after 8 P.M. This of course made it impossible for him to carry on his work at Mount Wilson. Only the intercession of powerful allies at the observatory, who insisted that Baade was not a German agent, enabled him to get his nighttime privileges back.[48]

Osterbrock defended him against the charge that Zwicky made to Baade's face, calling him a Nazi.[49] To Osterbrock, this was more proof of Zwicky's difficult personality.

There is no record that Baade ever joined the Nazi party, but there is one curious episode that might have played a part in Zwicky's accusation. It occurred in 1937, when Hitler was firmly in charge in Germany and girding for war. Baade was offered the directorship of the Hamburg Observatory. It was an appealing offer. When Baade left Hamburg, he had been a lowly assistant; the idea of returning as director of one of the most prestigious observatories in Germany was tantalizing. He even wrote that he was sure he could work with the Hitler regime.

In the end, Baade turned down the job. He realized that his career would not be advanced by leaving the greatest observatory in the world, even for the top job at a lesser institution. Work was under way at Palomar, on the 200-inch Big Eye, the dome of which he described as "like a gigantic cathedral." Imagining all the strange and wonderful things it would discover thrilled him. But that didn't prevent him from using the Hamburg job offer to pry a $2,000-a-year raise from the director at Mount Wilson, Walter Adams.

In his final letter to the Hamburg Education Board, withdrawing his name from consideration, Baade signed, "Heil Hitler."[50] It was a strange act for a man who no longer had any reason to curry favor with the government in power, since he was refusing the job. But it was very much in keeping for a man who remained a German sympathizer even after Hitler's troops invaded Poland in 1939.

The astronomer Allan Sandage, who knew both men well and admired Baade, said the men's relationship soured in the forties due to Zwicky's "unreasonable claims of having solved certain supernova problems in advance of Baade." Of course, Sandage, in the sixties, was embroiled in a nasty, and very public, feud himself with Zwicky, which no doubt colored his opinion.[51] Certainly, Zwicky could be difficult. The question is why.

Osterbrock reflected the clubby world of astronomy by deriding him as a "beginner in astronomy" who was "highly insecure." Zwicky was indeed self-trained, and his working method was far different from the average stargazers who were already coming to dominate the field. In the coming decades, as he increasingly felt his isolation, he would become angrier and angrier at the "grey thinkers" who guided their lives, and work, according to the "voodooism" of accuracy, while making sure lone wolves with inspired ideas were shunned.

Whatever the basis of Zwicky's and Baade's feud—and it was likely that there were several causes, as partnerships among powerful and innovative people are not known for their longevity—their enmity would be lasting and only get more bitter. "They became a dangerous pair to put in the same room," according to Jesse L. Greenstein, who supervised Zwicky, with varying degrees of success, as chair of Caltech's Astronomy Department from 1948 to 1972.[52]

Stories circulated that Zwicky threatened to kill Baade. It is doubtful that the Swiss did any such thing. There were stories of other threats which, when tracked to the people who supposedly received them, were promptly denied. One colorful story held that when Zwicky discovered a rattlesnake in the dome of the Schmidt telescope where he hunted for supernovae, he accused Baade of planting it.

Whatever the facts, the end of their friendship concluded one of the more fruitful working relationships in astrophysics. Both men would go on to make discoveries that would shape our understanding of humanity's place among the stars. Baade would do as much as anyone to show the true size of the universe and to break down the genomes of stars. Zwicky would make contributions in jet propulsion and rocketry that would aid America's war effort and get the country off the starting line in the space race.

But the breach would never heal for Zwicky. The fierce individuality that drove the two men apart, the boldness, confidence, and willingness to go his own way, would become as much a fixed part of Zwicky's working method as his insights. As would his attacks on colleagues, which in turn increased his isolation and estrangement from mainstream science. He claimed not to care, insisting that he had no desire to be liked by the charlatans he felt filled academic ranks. On rare occasions, and never in a public setting, this bravado would crack, just a little. "I have grumbled and cursed enough all these years, I think it was necessary, but it often torments me nevertheless," he wrote several years later to his friend Rösli Streiff.[53] In another morose note, he acknowledged that whenever he tried to force solutions, "it has always gone wrong."[54]

In the late 1930s, however, the Swiss was still relatively young, raw, and at the peak of his powers of inspiration. For now, the critics stood silently by, intimidated by the boldness and accuracy of his vision. When he presented photographs in December 1939, that he said proved the universe could not be "blowing up," there was no criticism from his colleagues at Caltech. He was too formidable a presence.[55]

The wider public, which would never really forsake him, considered him one of the great men who would lead the way to the utopian future in which science would solve all the nasty old problems of war and poverty that had afflicted human society since the beginning of time. This was how Fritz Zwicky saw himself and how he wanted to be remembered.

As his decade of discovery drew to a close, he was not just in the funny papers but the subject of poetic paeans. An amateur poet named Dorothy Kiel penned a simple ode to "F. Zwicky, supernova hunter of Palomar Mountain."

"Twinkle, twinkle, little star; soon we will know who you are. If you haven't got a name, You will be supplied with same. Just explode, do something tricky. Catch the eye of Mr. Zwicky! Don't forget, though, little star—Do it over Palomar."[56]

✳ 7 ✳

TILTING AT WINDMILLS

AS WALTER BAADE set sail for New York after the Paris colloquium, where his and Zwicky's work had been examined, debated, and ultimately saluted, Zwicky's brother-in-law, Nick Roosevelt, was fretting over the dangerous course of events in Europe. "It's all a horrid mess," he wrote to Fritz and Dorothy on August 21, 1939.[1]

When Britain's prime minister, Neville Chamberlain, returned from Berlin the previous September, claiming that he brought "peace for our time"—at the expense of the Czechoslovakians—civilized people hoped that Hitler's appetite for villainy had been satisfied. Instead, it had only been whetted. As Nick sent off his letter, German armies massed on the Polish border; Hitler was secretly negotiating to divide the spoils of Europe with Stalin's Soviet Union. War was coming, even if Chamberlain could not see it.

To Nick, working as an editorial writer at the *New York Times*, these events showed it was time for his distant relative, the current US president, to declare America's unwavering support for the European democracies in the hope of convincing Hitler to back off. FDR wanted to do just that. Unfortunately, his hands were tied by Congress, which had passed the Neutrality Act of 1935 to prevent America from getting involved in any foreign conflict. Now, with the drums of war beating, the act had been amended to allow arms to be purchased by foreign governments, but only if they supplied their own ships to carry them. This was the "cash-and-carry" neutrality Baade raged against. Like many, he saw it as a cynical betrayal of America's professed principles. The Americans might not want to dirty their hands in foreign conflicts, but they were happy enough to supply the weaponry for others to bloody themselves. But only if the buyers had ready cash.

After German troops marched into Czechoslovakia in September 1938, FDR sent his secretary of state, Cordell Hull, to Capitol Hill to ask Congress to scrap the neutrality laws altogether. Congress refused, handing President Roosevelt a serious defeat.

"I did all I could to support Mr. Hull and got a lot of criticism from Republicans for so doing," wrote Nick. Like most Republicans, Nick was no fan of the current occupant of the White House, despite their family connection. As he battled the Great Depression, FDR was bringing about the most radical transformation in history of the role of the executive branch of government, ramming through social programs to put Americans back to work and trying to pack the Supreme Court with his allies. Republicans were outraged at his presumption. The president had no interest in their opinions, they complained. He expected them to merely rubber-stamp his policies. "To us this sounded like a precept taken from the practice of Mussolini or Hitler," Nick Roosevelt wrote later.[2]

When it came to amending the neutrality laws, though, he and FDR were in accord. Both wanted to remove the congressional shackles. But Democrats controlled Congress, and they were in an isolationist mood. Then, when it came to a vote, Republicans joined them, more to frustrate the president than to support the opposition party. "I am sorry they voted as they did," Nick wrote to Zwicky and Dorothy.[3] "I did what I could to stop them, but, after all, even though I am Tirzah's husband I am but a poor feak and weeble [sic] creature with no real control over anyone (including Congressmen and Tirzah) and only able to say what I think and—in the case of Congress—say it in print, which I did. Had not many westerners been—I think mistakenly, but nonetheless sincerely—opposed to the changes which Mr. Hull wanted, those changes would, of course, have been passed."

At the time, the western United States still remembered its frontier past. They had even less interest in making cause with far-off European nations than the cosmopolitans in the east.

Nick's best remaining hope for avoiding war was that the British had learned "that appeasement has failed as a policy." Perhaps, he said, the British would make it clear to Hitler that they would not sit back while he swallowed eastern Europe. But he didn't trust Chamberlain. "I certainly wish that old goat would get out, and that in his place would be put someone like Winston Churchill." He would get his wish, but not before Germany plunged Europe, and then the world, into deadly conflict.

"Doubtless to you both"—Fritz and Dorothy—"I will seem hopelessly dumb," Nick continued, "but I can't get over my feeling that, despite the crisis now coming to a head, war will be avoided and Mr. Hitler's ambitions will receive their first real check. . . . If, of course, Fritz is right and the British once more sell out, then war is inevitable."

As Nick noted, Zwicky didn't trust the British, an opinion he shared with his father. "On the whole, the English have always been selfish," Fridolin wrote as war loomed.[4]

His son had other reasons for cynicism. A native of eastern Europe, his childhood had been punctuated by repeated conflict in the Balkans. For him, the bloody trench warfare of World War I was not a chapter in a history book. He lived through it. In relative safety in Switzerland, perhaps, but close enough, metaphorically at least, to hear the sound of the guns and the moans of the dying.

And he was, of course, correct. Even before Nick's letter arrived in California, Hitler signed his nonaggression pact with the Soviet Union. A week later, on September 1, 1939, German armies invaded Poland. France and Britain upheld their treaty obligations and declared war on Germany. By the spring of 1940, Germany's Blitzkrieg was sweeping through Belgium and France. By then, President Roosevelt had at last succeeded in changing the neutrality laws; it was too late, war was at hand, and there would be no turning back.

Zwicky got no satisfaction from being right about the course of events in Europe. In a despairing letter to Rösli, he declared 1940 the worst year "in the history of the world."[5]

Even after all that, however, the isolationist movement in the United States remained strong, with a consequence that would soil the reputations of many of Zwicky's colleagues in the scientific world. The American Association of Scientific Workers, made up of hundreds of the most prominent scientists in the country, chose that critical moment, in May 1940, to issue its infamous peace resolution.[6]

"Science is creative, not wasteful or destructive," the resolution stated. Yet "the same scientific advances which have contributed so immensely to the well-being of humanity are made to serve also in increasing the horrors of war. The present conflict in Europe focuses attention on this perversion of science."

"The futility of war is especially clear to scientists, for war, as a method of solving human problems, is out of harmony with the rational spirit and objective methods of science," the resolution continued. It recommended unceasing efforts to "preserve peace for the United States."

Stripped of its high-minded language, the resolution was a bald appeal to the president to let Europe settle its own brawl. This attitude reflected a certain truth on the ground, as Italy's Benito Mussolini laid it out for Nick in an interview. "Twice in the course of our talk he warned me that Americans did not and could not realize how deep-seated were European hatreds," Nick wrote.[7]

The peace resolution was signed by five hundred of the nation's leading researchers and theorists. Zwicky was outraged at the perfidy of scientists who should have known better. He saw the resolution as a domestic Communist plot to prevent the United States from going to war with the Soviet Union. He was not alone. Within weeks, the *New York Times* wrote that some researchers at Harvard, MIT, and Tufts had resigned from the scientific association over allegations that a cabal of Communist sympathizers had pushed through the resolution.

Supporters of the resolution denied it. "The charge of Communist sympathizers . . . is not a true statement," said Kirtley F. Mather, a Harvard geology professor. "Many of our members are liberals and some are to the left of center, but they are not 'fellow travelers,'" he said, invoking a popular term for Communist sympathizers. "The peace resolution coming at the time of the Nazi attack brought about the dissension. It looks to me like a case of war hysteria."

Feeling helpless and angry that his colleagues could not see how much their interests, and their way of life, were at stake, Zwicky dashed off a bitter letter to his friend, Jan Oort, the Dutch scientist who pioneered radio astronomy.[8] Zwicky said that he wanted to return to Switzerland to volunteer for the nation's defense forces, but the Swiss government dissuaded him. Whether it was because it valued his work abroad or because the Swiss were unsure how much value a 42-year-old college professor would be to their national defense, was unclear.

"Over there everyone can at least do something to prepare themselves for the day when we will have to defend our independence," Zwicky wrote to his friend in Switzerland. "Here, one is rather helpless, all the more so as the Americans are inexplicably divided in their opinions. Whatever decision they make, they immediately develop a fear that they have become the victim of some foreign propaganda."[9]

As he often did in such cases, Zwicky reduced the debate to a personal fight. He singled out J. Robert Oppenheimer and A. H. Compton, two of the best-known scientists of the day and prominent signatories of the peace resolution, for special enmity. He sneered that those men added their names to a document urging "brother scientists against putting their knowledge in the service of military preparedness . . . at a time when England and Greece stood alone against the Nazis."[10]

He was still seething years later, when he noted that these same men "were to add to the 'perversion' (of science) by recommending . . . the use of the most hideous weapon ever spawned by scientific minds."[11] This was the atomic bomb, dropped on Hiroshima and Nagasaki, which Oppenheimer and Compton played prominent roles in developing. Zwicky, on a special secret mission at the close of the war, would be one of the first Westerners to witness the devastation wrought by the bomb. For him, Oppenheimer and Compton were doubly guilty, first by neglecting to oppose Hitler and his allies, the Japanese, when they might have been stopped, and second, by recommending the use of the terrible weapon to win the war.

Zwicky's family in eastern Europe was once again forced to deal with conflict, virtually on their doorstep. Fridolin wrote in the spring of 1940 that even though Hitler seemed preoccupied with western Europe, Switzerland and Bulgaria were "not out of danger." The Swiss declared their neutrality, but they were unsure Hitler would respect it. The government began making defensive preparations. In the Glarus valley, tank revetments were going up.

"Maybe I am being optimistic," Fridolin wrote, echoing Nick Roosevelt's hopes. But he believed it best to remain calm "and not paint the

Devil on the wall. . . . We will soon hear if sanity prevails or whether total war engulfs the world."[12]

Bulgaria also hoped to steer clear of the conflict, declaring its neutrality. At the same time, it embarked on a dangerous game with Germany. It hoped to recover territory it lost, South Dobruja, in the second Balkan war that had disrupted Fritz's childhood. In this it succeeded, through an agreement with the Axis powers. But he who would dine with the devil must have a long spoon. When Italy's invasion of Greece stalled, Germany demanded that Bulgaria allow its troops passage so that Hitler's army could relieve the Italians. Bulgaria was forced to conclude a treaty with Germany, which in turn exposed its cities to Allied bombing from the air. At the end of the war, Bulgaria changed sides again, joining the allied nations against Germany. That didn't help them when the Soviets arrived and installed a Communist regime that lasted until the breakup of the Soviet Union in 1991.

The Zwicky family would lose almost everything, including the lead shot factory and the family home in Varna, which Fridolin built with his own hands. When they finally got it back, it was a shambles.

Fridolin concluded his letter with a note about a newspaper clipping he had received from a relative in Vienna. "In the Palomar mountains, where the biggest telescope in the world is being built, Dr. Fritz Zwicky, with the German astronomer Dr. Baade, are exploring the brightest stars of all," the article said.[13]

By this time, Baade and Zwicky were likely not doing anything together, except wishing each other's destruction. But the news of their fractured relationship wouldn't become public for some time. Fridolin was as surprised as Rösli had been to discover that his son was now a prominent astrophysicist. He had not been one to boast of his accomplishments and discoveries to his family, in part because he feared they wouldn't understand them—and in this he was right. News of his son's fame in America reached his father only in dribs and drabs, rarely from his son.

Fritz Zwicky was never satisfied with complaining about things and waiting for others to set them right. For good, or ill, he was a man who acted on his beliefs. Over the years, he was often heard to say that if a man could do nothing else, he could at least write a letter. During World War II, he did far more than that. If the Americans couldn't understand how much their own interests were at stake in the war abroad, he felt it was up to him to do something about it. In May 1940, as the peace resolution reverberated through the scientific community, and as Hitler's Blitzkrieg overwhelmed France, Zwicky launched a one-man campaign to interest the US government in an idea that he was so sure would succeed against the German armies that he boasted to friends that Hitler could be stopped with fewer than 100,000 men.

He first turned to his brother-in-law, Nick Roosevelt, who introduced him to Bernard Baruch, an influential businessman and philanthropist, who was a close confidante of President Roosevelt. Raised in South Carolina and New York, Baruch was the son of a physician who made his fortune on the stock exchange before the age of 30, speculating in the sugar market. Baruch was the perfect contact, not only because he was a friend of the president, but also because he served in a unique capacity as a special adviser to the Office of War Mobilization. In that role, he could cut bureaucratic red tape to work directly with industrialists and business leaders to determine what arms and supplies were needed and who should make them.

"Dr. Zwicky, who is a Swiss citizen married to an American (who happens to be my wife's sister) has for months been working on a project which may be distinctly useful in our preparedness for war," Nick wrote to Baruch.[14]

Because of the sensitive nature of the idea, he declined to elaborate on it. "Perhaps the best measure I can give you of the importance which (Zwicky) attaches to the plan lies in the fact that even though he is in the midst of serious research experiments in conjunction with the supernovae that he has been studying at Palomar, he is prepared to take as much time off as necessary" to work on the plan. "The value of the project can best be determined by a few discrete persons who, like yourself, can judge the technical problems involved."

Baruch was receptive. Despite his wealth, he prided himself on a common touch. In the New Deal, informality was in, and making cause with the working class was considered a mark of good breeding. In Hollywood, screwball comedies poking fun at the rich were the rage. At FDR's White House, Nick Roosevelt noticed, when he visited and took a swim with the president, police guards slouched with their jackets unbuttoned. Baruch, according to the fashion, became known as a "park bench statesman," for his habit of sitting on a bench in Lafayette Park in

Washington, DC, where he would discuss politics with any citizen who dropped by.

Zwicky traveled east in June to meet with Baruch and outline his plan, which he assured him would provide the democracies with a new and revolutionary offensive weapon that would exploit the inherent weaknesses of the totalitarian states. Baruch was impressed enough to pass him on to General Frank McCoy, president of the prestigious Foreign Policy Association, for an assessment of the soundness of the idea. McCoy, in turn, asked for a more complete plan to be worked out with skilled technicians.

By mid-July, the German invasion of western Europe was well under way, and the Soviet Union was engineering a series of coups in the Balkans as part of its secret agreement to carve up Europe with Germany. In his anxiety, Zwicky's bluntness surfaced in a way that would imperil any chances he might have had to influence policymakers in Washington.

In a letter to McCoy, he thanked him for his help and recounted his most recent meeting with Baruch. At that meeting, Baruch, according to Zwicky, "had various objections. Although he seemed to me somewhat incoherent, I gathered that he did not think it his business to dally as a private citizen with plans such as I suggested."[15]

On the contrary, dallying in military affairs was exactly Baruch's business. More likely, he was not convinced of the usefulness of Zwicky's idea, which, as usual, was based more on his own insight and gut feeling than on any familiarity with strategy and ordnance. It's not hard to imagine Baruch being taken aback by his visitor's passion and persistence.

Having hit a dead end, Zwicky did not retire from the field, but pressed on until, by sheer force of personality, he managed to get an introduction to Henry Wallace. At the time, Wallace was Secretary of Agriculture and, despite his membership in the opposition party, FDR's vice-presidential running mate for his unprecedented third campaign, in 1940.

Whatever else one could say about the professor from Caltech, he showed what one person with a strong vision and a stronger conviction could achieve in an entirely unfamiliar field, in this case politics and warfare. Zwicky had a long talk with Wallace, who sent him to Harry Hopkins. Perhaps Roosevelt's closest adviser during the war, Hopkins virtually lived in an upstairs bedroom in the White House. The man who supervised the Lend-Lease program, Hopkins, as much as anyone, kept

Britain supplied with the ships and armaments it needed to hold off the Germans until the tide of war turned.

As July passed into August 1940, Zwicky kept up his frantic schedule of meetings, shuttling from Washington to New York and back to California. He met with the Chinese ambassador to Washington, Hu Shih, to try to convince him that his idea might help them fight off the Japanese. He also met with Vannevar Bush, head of the Office of Scientific Research and Development, which would supervise the Manhattan Project that built the atomic bomb. Then came R. H. Fowler, an Englishman attached to the National Research Council of Canada.

In September 1940, as Germany consolidated its victory over France and Italy prepared to invade Egypt, Fowler suggested that Zwicky take his idea directly to the nation that might best be able to put it into practice—Great Britain.

At the time, the battle in the skies over Britain had been raging for two months. The Germans had hoped the defeat of France would lead the British to sue for peace. Instead, Winston Churchill, having ascended to the prime minister's office, as Nick Roosevelt had hoped, made one of his famously defiant speeches. "The Battle of France is over," he growled. "The Battle of Britain is about to begin."

At the height of the air battle, as the Luftwaffe rained fire on London, Zwicky got in touch with Professor Patrick Blackett, who had just been made scientific adviser to the British Anti-Aircraft Command. Zwicky told Blackett that he had long been interested in "various types of combat strategies" and had some ideas the British might benefit from.[16] Noting that others had turned him down "on various grounds," which he said subsequent events showed to be incorrect, he was turning to Britain. Now, for the first time, he put his ideas down on paper in detail. After extensive analysis he had come to believe that "the most effective combat strategy would be what may be called a three-dimensional war." The weapon with which he proposed to wage this war was . . . the helicopter. At the time, autogyros were new on the scene. In Zwicky's mind, the hovering ability of the "windmill machines" would allow them to drop bombs more accurately than speeding bombers.

He envisioned squadrons of copters swooping in over battlefields, knocking out tanks and supply depots, and whooshing off before the enemy could respond. "For night defense, so necessary now for big towns like London, the windmills could constantly be hovering at great altitudes with one or several 'piano wires,' two to three miles in length, suspended from them." A bomber running into such a wire at night would be disabled. "In view of the lack of any effective defense against night bombing . . . the suggested wire curtain might be very worth while."

At sea, helicopters would have the advantage of not needing a long runway, so more of them could be crowded onto the deck of an aircraft carrier.

He closed with an expression of "admiration for the magnificent fight which you and your great people are putting up in behalf of human freedom. Many of us feel ashamed that we are not off hand in a position to do more to help you."

In this one letter, and in the episode as a whole, Zwicky displayed everything that made him such a confounding and tantalizing figure: the insight, the fearlessness to embrace unconventional ideas, and the quixotic belief that in the midst of a life-and-death struggle a nation could reboot its defense industry to employ an untried weapon. Wire curtain, indeed. Another twenty years, and another war, would pass before the helicopter would assume the combat role that Zwicky envisioned. In Vietnam, autogyros, by then known as Huey attack helicopters, equipped with rockets and .30 caliber machine guns, would become a key battlefield weapon in countless jungle conflicts.

Blackett's response was a classic in British understatement. He said it would take years of research and development to make the autogyro an effective weapon, then another three years to build "large numbers of such machines. . . . At the moment we cannot take such long views . . . and must stick closer to conventional design."[17]

And with that, Mr. Zwicky's first foray into tactics and armaments came to an end, not with a whimper or a shout but with a polite but firm kiss off. His frenetic efforts in the summer of 1940 had come to nothing. It turned out to be an unrealistic flight of fancy to think a college professor from Switzerland, by way of California, could influence events in a worldwide war. And yet, the next time he tried his hand at war making, he would do exactly that.

✳ 8 ✳

ROCKET MAN

IN JANUARY 1945, a Navy Liberator was returning from bombing runs against Japanese forces in the South Pacific when the aircraft developed engine trouble.[1] The B-24 was a notoriously unreliable aircraft, with a distressing tendency to catch fire in flight. On this occasion, the crew—eight airmen and a pet dog—tried to nurse the plane along, but they were 500 miles from the California coast. After radioing their location, they put the big bomber into the water and scrambled into a life raft.

Bobbing on a flimsy raft in eighteen-foot seas, their chances of survival were not good. Three hours after splashing down, the stranded crew's prayers seemed to be answered when they saw the sun glinting off the silver bodies of American patrol planes.

The planes circled, but the rescuers could do little more than wave at their comrades, soaked and shivering in the water below. The swells were so high that conventional seaplanes could not set down. If they tried it, they'd likely never get up again. The number of stranded Americans would be that much larger.

Surface ships could not help. The nearest vessel was 350 miles away. Unthinkable as it seemed, there was every chance the men would drown under the watching eyes of helpless rescuers. As one report put it, only a miracle could save the airmen "before the angry sea battered their fragile raft to bits."[2]

That miracle appeared in the form of a Catalina flying boat under the command of Donald MacDiarmid, the ranking officer at the Coast Guard station in San Diego. Defying the waves that kept the other planes on their wings, MacDiarmid slammed his ship into the water. The landing was rough, but the plane stayed up, and his crew pulled the exhausted Liberator men—and their mascot—aboard.

That was the easy part. Getting out would prove whether this was a rescue mission or a bit of reckless folly. Waves sloshed over the engines as MacDiarmid revved up. The Catalina's four big propellers beat the water and the plane plowed sluggishly forward as the men looked anxiously at each other. Suddenly, there was a whoosh and trails of smoke shot out from each side of the plane's hull. The craft leaped forward and up, surging out of the water at an unaccountably steep angle. Everyone came home that day, including the dog.

"We'd never have made it without the jatos," Commander MacDiarmid said. "Jatos get you up in a hurry."

"Jato" stands for jet-assisted takeoff. It was a brand-new wrinkle in aviation. As America carried the war deeper into the Pacific Ocean, and into Japan's backyard, the jato played an important part in the Navy's success. Not only did the small jets enable seaplanes to rescue downed pilots—the Navy estimated they saved 2,870 lives during the war—but they also changed the way the war would be waged in the future.

As late as 1942, American aircraft carriers needed thirty to forty seconds to launch a single plane with catapults.[3] The so-called baby flattops were worse. They could launch planes only when the winds were just right. This made the carrier, in the words of a Navy report "a weapon that could not be used in the forward spearhead of a task force attack."[4] This fact dramatically compromised the lethal power of the carrier as the armed forces moved toward what was expected to be a very bloody assault on the Japanese homeland.

Essentially, the Navy admitted, its carriers were "tactically and operationally obsolete."

But with jatos strapped to their bodies, Navy fighters like the Grumman Wildcat and Avenger and the Vought Corsair could get airborne fast. By mid-1944, American carriers were launching coordinated, thousand-plane assaults, turning the tide of war in the Pacific decisively in favor of the Americans.

MacDiarmid didn't much care that day where the technology came from that allowed him to save fifteen lives, including his own. What mattered was that it worked. He had no way of knowing that those small jets strapped to his seaplane had a history that stretched back to a group of putterers at Caltech and an imaginative Swiss astrophysicist whose inventions would help lay the groundwork for America's postwar rocket and missile industries.

It all started when three young men approached Professor Theodore von Kármán for permission to experiment with rockets on the Caltech campus. At the time, in 1936, von Kármán was the director of the Guggenheim Aeronautical Laboratory at Caltech, known in military parlance as GALCIT. A former fighter pilot with the Hungarian military in World War I, von Kármán was one of the world's leading aerodynamicists.

Only one of the young men was a student at Caltech. Frank J. Malina had studied physics under Fritz Zwicky. By the time he and his two friends, Edward Forman and Jack Parsons, approached von Kármán, they had already been turned down by other members of the faculty. One instructor told Malina he should stop wasting time on "harebrained" ideas and get a job in the aircraft industry that was booming in southern California at the time.

Malina and his friends were uninterested in conventional aircraft work. They aimed much higher. They wanted to launch rockets that could breach the atmospheric envelope that encircled Earth and plunge into space. They told von Kármán they hoped to learn more about the mysterious cosmic rays, the source of which was still being hotly debated.

Unlike the more conservative-minded faculty, von Kármán, a lovable imp with a rapier sense of humor who favored berets and to the end of his days spoke with a thick accent, was captivated by the "earnestness and the enthusiasm of these young men."[5] Von Kármán's humor and willingness to go his own way, though without the brashness of his friend Zwicky, attracted so many imaginative young people over the years that a term was coined for them: the Kármán Circus. Confronted by this group of dreamers, he didn't hesitate to offer his support. He agreed to let the "young fellows" use the resources of the aeronautics laboratory after hours, making Caltech the first American university to treat rockets seriously. Even so, America was years behind Germany, but no one could know how much. The rocket research that the Germans were doing was shrouded in secrecy.

It was a testament to von Kármán's keen eye that he saw something in the three friends. They were a most unusual group. Malina was a Texan who came to Caltech on a fellowship. Forman was the son of an engineer who grew up in Pasadena, where he met Parsons. The Parsons family backyard served as their first testing ground; the scarred and cratered lawn at his home was proof of their ambitions, if not their wisdom.

Parsons was a true American original. As interested in black magic as he was in rockets, Parsons would befriend and study under the Englishman Aleister Crowley, the occultist, poet, and ceremonial magician, who proudly embraced his reputation as "the wickedest man in the world." Parsons also became a close friend of L. Ron Hubbard, the founder of Scientology. Over six feet tall, with wavy dark hair and a sly, knowing manner, Parsons was irresistible to women. All that mattered to von Kármán was that he also happened to be an excellent, if self-taught, chemist, who had many good ideas about how to make fuels capable of sending a rocket skyward.

The three young experimenters were certainly not the first to be interested in rockets. The rocket is an ancient invention. The Chinese used it in battle as early as the eleventh century to propel "fire bombs" at their enemies. In America, Robert Goddard had been experimenting with rockets since 1915 on the campus of Clark University in Worcester, Massachusetts, where he taught. In 1926, he fired the first liquid-fuel rocket, which reached a height of forty-two feet, making him the modern father of rocketry. When the German rocketeers were interviewed after the war about their rocket program, they said they learned everything from Goddard's writings. But Goddard himself was so suspicious of outsiders that his early discoveries played almost no role in the development of America's billion-dollar rocket industry.

It was left to the Californians to take what was then known about propulsion and harness it. With little money, the young experimenters scavenged steel tubing and other material from junkyards. After experimenting with different fuel combinations, by the beginning of 1937, they had made enough progress that von Kármán allowed them to keep going.

Problems developed along the way. On one occasion, a rocket misfired, sending noxious fumes throughout the building. Von Kármán got so many complaints that he moved the tinkerers outside to a concrete platform next to the aeronautics laboratory. One night, the building shook from two explosions. They were so violent that a measuring gauge was sent flying, lodging deeply into the building's wall. Malina was away at the time. If he had been seated at his usual stool, he would have been killed. The mishap led Caltech students to tag the group with their famous nickname, one that captured their daring, recklessness, and heedless enthusiasm: the Suicide Club.

After that accident, von Kármán sent them off campus, to the Arroyo Seco Creek bed, behind Devil's Gate Dam. There they set up their test stands and kept at it. After another year and scant progress, Malina went to see Goddard in Roswell, New Mexico, where he was working at the time, in hopes of winning his help. No such luck. When he wasn't sure others were stealing his ideas, Goddard was certain he was being ridiculed. He carried around a newspaper clipping from the *New York Times* that made fun of rockets, apparently to remind himself of the danger of trusting outsiders. Ridicule was not an uncommon reaction at the time, even among scientists who should have known better. MIT refused government funds to study rocket technology, considering it "Buck Rogers stuff." The term was so out of favor that von Kármán and his young men decided not to call their work a rocket program.

The military was not interested in rockets, either. But they *were* interested in anything that could get airplanes into the air faster. With war looming and the big American aircraft manufacturers like Boeing, Consolidated, and Douglas designing giant bombers, the military knew those big planes would need ever-longer runways, unless there was some other way to get them up. The answer was the jato.

Jet propulsion is now so familiar that it falls into the category of the light bulb and the radio, devices so common that people think they know how they work, even if they don't. And if they don't, it doesn't matter so long as the devices operate as expected. But in Zwicky's day, the idea of jets was strange and new.

There are a wide variety of jet engines, but they all operate on a simple principle: An explosion releases gases that propel the vehicle forward. In engineering terms, it's called "reaction propulsion," a term that deeply upset a reporter from behind the Iron Curtain who interviewed von Kármán years later. "Please," she said, "how can I say in a progressive paper that progress is accomplished by reaction?"[6]

A fundamental difference in the various engine types is the source of oxygen that fires the engine. Rocket motors, like the fearsome German V-2, carried their own supply, while airstream engines, like the ramjet and pulse-jet, get their oxygen directly from the atmosphere. In the rocket, the oxygen that enables the fuel to ignite is furnished by a compound called an oxidizer. The oxidizer is mixed with the fuel—together they are called propellants—which can be any hydrocarbon, as well as liquid hydrogen. The mixture is then ignited, sending the vehicle hurtling skyward and, eventually, spaceward.

In a pulse-jet, compressed air is forced into a chamber. When the pressure reaches a certain level, a flap opens and the compressed air enters a combustion chamber, where it is mixed with gasoline or some other fuel and fired. The blast blows the flaps closed again, and the process repeats. The German V-1 "buzzbomb" that was fired on Britain in 1944 used this motor. It was the rat-a-tat sound of the flaps, opening and closing, that gave the buzzbomb its chilling nickname. The Germans had another name for it, *Kirschkern,* or cherry pit, which they spat at London by the hundreds.

The American military didn't know what the Germans were up to, but they did understand American aviation technology was out of date. They thought there might be a use for the little rockets von Kármán's young men were setting off in the creek outside Pasadena.

The Caltech crew got their first big contract from the military, $10,000, in July 1939, two months before Hitler invaded Poland. The war was on, but the young rocketeers were very far from finding a reliable propellant that could help planes launch from carrier decks, let alone pierce space. Finally, the suiciders came up with a solid propellant made of roofing tar and potassium perchlorate. Parsons got the idea while watching a construction crew mix molten asphalt.

Solid fuels were powders, or cakes, loaded into narrow steel tubes about the size of a fire extinguisher. Once ignited, they burned something like a cigarette, from one end to the other. The trick was slowing the burning sufficiently so that they could burn long enough to get a plane airborne. Around twelve seconds was considered ideal.

Liquid fuels operated differently. They comprised a fuel and an oxidizer stored in separate tanks. When injected into the combustion chamber, they spontaneously ignited. In some ways, liquid fuels were superior. They produced a higher specific impulse, the technical term for thrust, and had the advantage of being able to be turned on and off. But they were less reliable and more difficult to handle, since the chemicals were often toxic and very acidic.

For almost two years, Malina labored on the fuel problems, explosions regularly echoing out of the canyon west of Pasadena. The site would eventually become home to NASA's Jet Propulsion Laboratory, but at the time, it was sufficiently remote that the authorities received few complaints about the Suicide Club's activities.

Just three months after the Japanese attacked Pearl Harbor in December 1941, Malina, working with Martin Summerfield, came up with the idea for a new liquid fuel, using aniline and nitric acid. According to von Kármán, "the results were spectacular."[7]

On the basis of field tests, General H. H. "Hap" Arnold of the Army Air Force—there was as yet no independent US Air Force—urged von Kármán and his suiciders to go into production. The whole crew met at von Kármán's home, each man putting up $200 to get things going. But that was never enough. Soon, they were dipping into bank accounts, taking second mortgages on their homes, and leveraging even the cars they drove.

In March 1942, they opened for business in the back room of a Vita-Juice health food store. They named their new company Aerojet. They had tossed around other names. Von Kármán threw out Superpower, but that was rejected. Anything with the word rocket in it was not even worth discussing. So it was Aerojet. The six-man company, which would eventually be known as the General Motors of rocketry, barely registered as a blip on the radar screen, even for the armed services that inspired its creation.

Their first significant order came from the Army Air Force, which contracted for sixty liquid propellant motors, telling the little company to stand by for a larger, $200,000 order. But then, with their homes and cars, everything they owned, mortgaged to the hilt, the Aerojet crew was told that the Army had changed its mind. They didn't need jatos after all; Army engineers had become so proficient at building runways that the brass decided it would be easier for the bulldozers to build longer airstrips than to equip thousands of planes with jet packs.

As von Kármán put it later, "the bulldozers were moving faster than our research."

The little Pasadena company was out of business, another casualty of war, before it even got off the ground. Just at that moment, the Navy came knocking. As the Army's difficulties were easing, the Navy's were increasing. Carriers were ranging deeper into the Pacific. In August 1942, marines landed on Guadalcanal and took the airfield. The Japanese waged a bloody, months-long assault to retake the strategic island.

To protect the men fighting and dying in the Pacific, and to rescue those shot down or stranded in the ocean after their submarines sank, the Navy needed to get more planes aloft, faster and with heavier loads. Engineers couldn't make carrier decks longer, so the Navy looked for another answer. Aerojet's jatos seemed like the solution. The trouble was, Aerojet was again having trouble with its propellants.

An embarrassing test before the Navy's top brass, aboard the aircraft carrier USS Charger, out of Norfolk, Virginia, demonstrated the problem. A Grumman fighter, with Aerojet's solid rockets attached, took off with a whoosh. The test seemed like a big success. Unfortunately, the jet packs left behind a big cloud of smoke, which coated the officers' uniforms with a sickly yellow residue. One officer stalked off, saying the device "was impossible."

It was obvious to everyone that leaving behind a cloud of yellow smoke every time you launched a fighter was a bad idea. "These mushrooms standing over the ocean like little mountains," in Fritz Zwicky's words, would betray the carrier's position to hunting Japanese patrol planes.[8]

The Navy told the Aerojet crew to come back when they had solved the problem. This was pretty much how things stood when Fritz Zwicky was invited to a fateful dinner at von Kármán's home in October 1942. Zwicky and von Kármán were old friends. When Robert Millikan floated the idea of bringing the Hungarian aerodynamicist to Caltech, Zwicky had enthusiastically offered his support. After von Kármán's arrival in California in 1930, the two men resumed a friendship they had fashioned in Europe. Von Kármán, seventeen years older, tended to look after his Swiss friend the way an older brother would a talented but headstrong sibling.

Von Kármán's good humor and quiet self-confidence allowed him to see his Swiss friend's gifts without being threatened by his brashness. That didn't mean he was blind to his friend's peccadilloes. He once told Fritz that he was thinking of coining a term for the roughness of an aerodynamic surface. He would call it a Zwicky, he said. But then, he realized that nothing could measure up to the boldness of a full Zwicky, but the man himself "so the practical unit will be a micro-Zwicky."[9]

Von Kármán told the story many times, and Zwicky referred to it on occasion. He wasn't convinced it was as funny as von Kármán thought it was, but the diminutive Hungarian had such an eye-twinkling way of having fun with you that it was impossible to be angry.

Also at the von Kármán dinner table that night in October was Andrew G. Haley, a lawyer who had just been appointed president of the Aerojet Engineering Corporation. Von Kármán was the company's first president, but when orders for jet motors started coming in, he stepped aside for Haley, who had represented corporate interests in the radio industry and understood what was necessary to run a business. Later on, he would help invent the field of space law, drawing up many of the conventions that govern the use and exploitation of space today.

The von Kármán household on Marengo Avenue, one of Pasadena's older neighborhoods, was a welcoming place without undue pretension. The house was decorated with tasteful but unostentatious Oriental furnishings; above the fireplace in the dining room was a large portrait of von Kármán's father, Maurice. His sister, Josephine, known as Pipa, presided with the ease and manners of a hostess in diplomatic Washington. Von Kármán, once described as "the gnome with the Mona Lisa smile"[10] for his gentle, self-mocking humor, supplied the lively conversation and gentle jibes at those who dared put on airs. When bored, rather than insult his guest by excusing himself, he surreptitiously turned off his hearing aid.

At dinner, Zwicky was charming as he described his work. Haley listened with fascination. He was particularly intrigued by the thought process Zwicky used to attack problems. What was that again? Morphology?

Zwicky loved talking about Morphology. He launched into a detailed description of how he approached every problem. Morphology was not just a habit of thinking, he said, it was "a way of life . . . attempting to realize the genius of each individual and each race." The prime directive, as he put it, was to "generalize all problems before drawing fallacious conclusions."[11] In practice, this meant keeping one's mind open to all possible solutions, no matter how seemingly impractical. Zwicky traced Morphology to Paracelsus, the sixteenth-century Renaissance physician, botanist, astrologer, and humanist. It was an apt comparison, since Paracelsus was a stubbornly independent reformer who disdained conventional wisdom governing the practice of medicine in favor of his own observations of nature. He was the first to argue that disease had a psychological component.

Zwicky noted with pride that Paracelsus was maligned in his own time. "The morphologist must accept such treatment," he would say, "because he will go into medicine, or into business, engineering, astronomy and so forth, and tell these guys about approaches, facts and concepts which they have forgotten. The specialists don't like that."

Zwicky was of course talking about himself and his struggles with colleagues at Mount Wilson and Palomar, and in science in general. He believed that "if the earth and humanity are going to survive at all, the next cultural style will be that of the age of morphology."[12]

Haley was bowled over by this idea, but even more so by the man who invented it. For a lawyer, Andy Haley was an excitable man, given to issuing passionate directives to his employees, exhorting them to "drive night and day with indomitable purpose" to get the job done. "This is the crossroads, with all our reputations at stake," he wrote in another dispatch.[13]

For such a man, the imaginative mind of Fritz Zwicky was irresistible. Aerojet was having trouble with its propellants, he said. Was it possible Fritz could come aboard to see if this morphological thinking of his could solve the problem?

Zwicky was, of course, flattered by Haley's proposal. But it was impossible, he said. He was simply too busy. Besides scouring the heavens for supernovae, he was just then managing civil defense for the city of Pasadena. Having seen the effects of gas attacks in World War I, Zwicky felt Americans were not sufficiently alert to the danger posed by similar attacks in their country.

In hindsight, the possibility that Japan would gas mainland California seems silly. But in the anxious months after the attack on Pearl Harbor, the danger of attack seemed very real. California beachgoers were asked to scan the waves for submarine periscopes, and coastal cities imposed nightly blackouts. The dark skies over Los Angeles created the best conditions the astronomers at Mount Wilson would ever again see, greatly aiding the researches of Hubble and Baade.

Once again, rather than sit on the sidelines, Zwicky had jumped in with his usual energy to solve the problem. He met with Pasadena's Public Health Officer, Charles Arthur, who was quickly overwhelmed by his visitor's passionate insistence that Pasadena was hopelessly unprepared for gas attacks. Arthur accepted the professor's suggestion that the city build special decontamination trucks, which could be dispatched on a moment's notice to spray neutralizing chemicals in "infested" neighborhoods. Naturally, Zwicky went to work designing the special vehicles himself. In the cab of the decontamination truck, he installed a filter system that pumped in compressed air so that no gas could invade, allowing the driver to enter the most contaminated areas without fear. Zwicky also designed his own "scram" gas masks, consisting of five layers muslin and cheesecloth. Thousands were manufactured for distribution to the citizens of Pasadena. Zwicky tested them against everything from tear gas to deadly phosgene gas. A squad of eighty-five men was trained to spring into action the moment a gas bomb landed.[14]

It hardly needs to be said that the city of Pasadena was by leaps and bounds better prepared for gas attack than any other American city. In the end, however, only one citizen was sickened by poison gas, and that was Zwicky himself, during one of his tests.

As Haley and Zwicky chatted over dinner at von Kármán's that night in October, Germany had just completed its first successful test firing of the lethal V-2 rocket at Peenemünde, on the Baltic Sea island of Usedom. Three previous tests had failed. But on October 3, 1942, the twelve-ton rocket lifted off its pad and flew 125 miles over the conquered lands of northern Europe, reaching a speed of nearly 3,500 miles an hour.

The United States had nothing in the works, at Aerojet or anywhere else, that could compare to it. Each V-2 carried a one-ton payload of bombs. To the Allies' good fortune, the Germans would be slow to bring it into action, in part because Adolf Hitler had a dream that the rockets would never work. The first V-2 launches on targets didn't occur until September 1944. They killed 2,700 British citizens, but by then the tide of war had already turned decisively against Germany.

If Haley had known what was going on in Germany, he would have been even more insistent. As it was, Zwicky's refusal only stoked Haley's desire to bring him on board. For months after the October meeting, he implored and cajoled the Swiss professor. He didn't have to take a full-time position, Haley said. If he could just take a look around to see what they were doing wrong, it would be invaluable in the fight against the dictators. Such an appeal went straight to Zwicky's heart, though his own preferred term for the Axis powers was the "so-called new orders."

Finally, Zwicky agreed. He was installed as a consultant, but by May 1943, he was working full time for Aerojet, clocking sixty-seven hours a week, raising the question: When did he sleep?

After performing a morphological analysis of the infant field of jet propulsion, Zwicky told Haley that Aerojet should establish a research department to work on all phases of rocket and jet engines. Haley liked the idea so much that he appointed Zwicky chief of the new department.

In an early report, Zwicky laid out the governing ideology of his department: "The fundamental philosophy of the Research Department has been that free men in a democratic society are second to none in inventiveness and in the determination to realize fundamental inventions in practice for the benefit and welfare of human society."[15] His team would be guided by three principles. First, "to think clearly," which for Zwicky was the key to all invention. Second, to have "the courage of one's conviction regardless of difficulties." This of course didn't need to be said, except inasmuch as it was important to instill in those who worked under him, the chemists, engineers, draftsmen, and the like the attitude that Zwicky brought to bear on every aspect of life. Finally, those who worked in the department would be obligated to do everything in their power to "create the necessary material and spiritual conditions" to produce new inventions that would help free people succeed in the existential struggle against the forces of evil.

The language today sounds histrionic but it should be remembered that when Zwicky wrote these things, the struggle was still in doubt. There was a lot of defeatist thinking that the authoritarian governments might have the advantage in war because they could make decisions quickly, without having to consult with bothersome civil institutions like congresses and parliaments. They could also use slave labor, as the Germans did at the Peenemünde rocket complex, while Aerojet had to observe labor laws setting pay grades and workplace safety.

At the same time, Fritz Zwicky would have stated these principles in precisely the same way no matter what was going on in the wider world. More than most, he was always in an existential struggle to prove the value and worth of the individual. If he was his own best example of a free man, that didn't mean that he was unique. Everyone could be just as free and inventive, if they only did as he did and liberated themselves.

Once at work, he quickly developed a low opinion of Aerojet's capacity to fulfill its contracts. The Navy had just given the company its largest contract to date, $3 million for solid and liquid jatos. But after the embarrassing test aboard the USS Charger, that contract was in grave danger.

Haley was disconsolate. "Problems relative to solid propellant development have been incessant," he fumed in a classified report at the time.[16]

For Zwicky, the "smoky and toxic exhausts" were only part of the trouble. A third of the motors blew up on the test stands. They also had a very narrow temperature range of safe firing. This was almost as great a problem as the smoky exhaust. To be useful, a propellant must be able to work as well on a cold, foggy morning as in the heat of the day, with the Pacific Ocean sun beating down. One problem, Zwicky discovered, was that each batch of asphalt they got from Texaco or Union Oil was different, "and we had all this uncontrolled burning going on," he said.[17]

He was not much happier with the liquid propellant, which was difficult to handle and was both toxic and corrosive to metal combustion chambers.

While Zwicky was still assessing the scope of the job ahead of him, Army Ordnance sent a packet by secret courier to von Kármán. It contained aerial photographs taken by British intelligence of the French coast. The military wanted to know what he thought of them. Slipping the pictures out of the envelope, von Kármán saw an odd-looking set of structures, which he recognized immediately as a launch complex. It was big, bigger than anything Aerojet was using, or envisioned, complete with platforms and storage areas. He was at first shocked, then alarmed. In his hands he held "the first tangible proof that Germany was doing something with large and novel missiles."[18] In fact, these were V-1 buzz-bomb installations that would soon be raining destruction on Britain. He hurriedly sent off his response: Nazi science was far more advanced than anyone had imagined.

Suddenly, the work at Aerojet developed a new intensity and purpose. "We must work with the highest energy and indomitable courage to justify" the confidence the United States had placed in them, Haley wrote to his burgeoning work force.[19] With a $250,000 grant from the Army, work had started on the nation's first propellant plant, located in Azusa, twenty miles east of Pasadena. By the end of the war, the complex would consist of eighty buildings spread over fifty acres. Zwicky's research department comprised 4,800 square feet of organic and inorganic chemistry labs, a plastics lab, and a physics lab. Attached to the lab complex were ten solid-propellant firing pits, three liquid-propellant firing pits, and eleven firing stands with control rooms protected by concrete revetments.

By the summer of 1943, Zwicky's department was up and running, allowing him and his team of chemists, engineers, mechanics, and draftsmen to get down to the hard work of making trouble-free propellants. The first task was to reduce the weight of the propellant systems, a key factor in determining how much lift the rockets could provide. He investigated different metals and added liners to prevent the corrosive chemicals from eating away at the combustion chambers.

Next, he went to work on the solid propellant. His initial plan had been to develop new propellants, but the problems with the existing propellants, particularly the solid one, needed immediate attention. The problem was not only the different qualities of the asphalt, or roofing tar, but also the presence of potassium perchlorate. Besides the smoky exhaust, it is relatively inefficient.

R. J. Mill

B. L. Dorman

R. B. Young

W. E. Campbell

W. E. Zisch

W. L. Rogers

E. S. Reichard, Jr.

C. A. Gongwer

W. C. House

E. R. Roberts

R. L. Queen

L. M. King

J. S. Warfel

R. D. Geckler

P. P. Datner

K. F. Mundt

Gen. Omar N. Bradley   Dr. Fritz Zwicky   Dr. Clark B. Millikan   Rear Adm. Calvin M. Bolster   Dr. Theodore von Karman   Dr. Bruce H. Sage   Dan A. Kimball   Arthur H. Rude

The Technical Advisory Board at the Aerojet rocket works in Azusa, in 1954. At the head of the table is Zwicky's longtime friend and patron, Theodore von Kármán, who once joked that he was going to coin a term for the roughness of an aerodynamic surface: It would be a micro-Zwicky. Zwicky himself is in the second row, next to General Omar Bradley.

Zwicky experimented with many compounds. He even wrote to his friend Tadeusz Reichstein to see what he knew about explosives and fuels. At last, he hit on ammonium perchlorate. He combined this compound with an oxidizer in a rubbery binder that held the mixture together. The resulting propellant, called Paraplex, was highly efficient as well as elastic and stable, which prevented the compound from developing cracks in storage that could lead to accidental explosions. Best of all, it was smokeless. Paraplex was perfected too late for the war, but through 1958, Aerojet sold 500,000 propellant units for jet-assisted takeoff and, later, for large rocket boosters, earning hundreds of millions of dollars.

Zwicky and his team also extended the safe firing range to as low as −20 degrees and as high as 150 degrees, an improvement of 200 percent that would allow Aerojet's jatos to operate in every type of weather. By slowing the burn rate, he was able to extend the operating time of each jato to fifty seconds, enough time to get even the slowest, most lumbering planes off the ground or carrier deck. Zwicky also developed a method that allowed a pilot to turn off the propellant at will, if, for instance, a problem developed elsewhere on the plane that required aborting the takeoff.

Finally, he was able to boost the thrust of Aerojet's solid rockets to 100,000 pounds. That meant the military would need fewer jatos to launch each plane, increasing reliability and safety. Compared to the big million-pound rockets of today, 100,000 pounds may not sound like much, but it was a very big and important improvement over the twenty-five-pound versions Malina and Parsons were setting off in Arroyo Seco only a couple of years earlier. The Suicide Club would become famous, but it was Fritz Zwicky and his team who took their good ideas and made them work on an industrial scale.

According to George McRoberts, who had only recently graduated from Caltech when he went to work at Aerojet in April 1943, Zwicky's innovations "became the basis for much of the advancement of rocket science that took Aerojet from 1,000-pound jato units to the multi-million pound thrust units we know today."[20] He made those remarks shortly after Zwicky's death in 1974, when the American rocket program had just landed a dozen astronauts on the moon. The giant Saturn V rocket, the tallest, heaviest, and most powerful rocket ever put in operation, the rocket that sent the astronauts to the moon, was built on the pioneering work Aerojet and Zwicky's team did in the forties. Their work also paved the way for the intercontinental ballistic missiles that safeguarded America during the Cold War.

Fritz Zwicky's salary at Caltech in 1943 was $4,124.97. At Aerojet, he earned $6,400. It was a good salary for the time, but a pittance compared to Aerojet's future earnings.

Once the crisis with the solid propellants had been met, Zwicky turned his imagination in a new direction. Using his morphological method, which he sometimes called directed intuition, he investigated new kinds of jet engines. One was a pulse jet engine similar to the one Germany would soon launch from the installations on the French coast. One entirely new engine, a hydropulse motor for use undersea, was propelled by a sodium fuel, which reacted violently in the presence of water. The development of hydropulse engines would inaugurate an entirely new field of underwater jet propulsion. Finally, Zwicky started the infrared program at Aerojet, which was widely used for missile tracking.

All these things were new and untried. Finally, Zwicky and his team came up with the world's first monopropellant, in which the oxidizer was contained in the fuel. By the end of the war, Zwicky's name was attached to eighteen patents, for launch apparatuses, the hydropulse motor, solid and liquid propellants, and the monopropellant.

"I firmly believe that when the history of Aerojet is written he [Zwicky] will be regarded as the catalyst who brought together many of the more important elements that contributed to the company's success," von Kármán wrote later on.[21]

"To a group of young engineers, it was a minor miracle to be working so closely with Dr. Zwicky and other Caltech professors," McRoberts said. He was especially impressed by Zwicky's fearlessness when it came to new ideas. He was willing to try anything, literally, to see if it worked. This characteristic in a man in his mid-forties was remarkable to the younger man.

During Zwicky's tenure, the research department grew from a handful of white-coated lab technicians to 300 chemists, engineers, physicists, and draftsmen. All reported to Zwicky, who at the same time taught classes at Caltech and hunted stars at Palomar.

Zwicky's employees were almost all young, recent college graduates, like McRoberts. The Swiss always enjoyed the companionship of the young more than the so-called experts with fat resumes. A joke at Aerojet in those days held that unless you were wearing braces, you had no chance of landing a job in the company's research department.[22] When the workday was done, he often went skiing with his young charges, teaching them to manage difficult runs on Mount Baldy and elsewhere.

The tremendous growth of the company didn't come without costs. There were accidents, some of which claimed lives. "They all knew it was dangerous, but they knew the job had to be done," said Margaret Kelley, whose husband Richard died in an explosion not long after the end of the war.[23] A survivor said he was blown twenty-five feet over a revetment. "All they were doing was scraping out one of the mixing kettles," Margaret said, as neighbors gathered around. "Well it has to come some time for all of us." The accident was proof of how dangerous the work was and how delicate were the chemical reactions Zwicky's department was probing.

With the problems with the propellants seemingly in hand, it looked like the little company that started out in a fruit juice store was set for un-dreamed-of success. But just at that moment, the company encountered perhaps its greatest crisis. Haley called from Washington with the alarming news that the military had once again lost interest in rockets and jet propulsion. It was one of those paroxysms of official doubt that all government contractors are familiar with. The government man, or team, or branch that had been so enthusiastic had moved on, and the new guys decided that the original idea was a boondoggle. According to Haley, the very survival of Aerojet was at stake. The only person who could save the situation was the director of research. That in itself was extraordinary. Haley could find no one better to communicate the vision and future of jet propulsion than the guy in the lab coat with a piece of chalk in his pocket.

In January 1944, Zwicky flew to Washington, where he met with one general after another over a period of two weeks. Two full days were spent presenting his morphological view of the future of rocket power. While in Washington, he tried to interest the brass in a powerful new engine he called an aeroresonator—his pulse jet. Most of his listeners, in the words of a contemporary writer, "rose and snorted out of the room. Later that year the engine he had described began dropping on London in the form of the V-1 buzzbomb."[24]

Nevertheless, his trip was a success. Zwicky returned with a $350,000 contract that kept Aerojet afloat while it looked around for investors with deeper pockets. In fact, the military was very concerned about the financial and corporate structure of a company run by college professors and their students. One of Haley's clients at the law firm where he worked before joining Aerojet was General Tire and Rubber Co. of Akron, Ohio. Haley knew the company was looking for investments, so he approached the tire manufacturer. General Tire was indeed interested.

"Here was a collection of great talents sitting on a multi-million dollar backlog of jato orders from the Navy," von Kármán recalled.[25] Banks might not see value there, but the Akron company did. They let it be known that they were willing to offer a $500,000 line of credit. General Tire would also assume the burden of meeting contract deadlines, relieving Zwicky from making any more emergency trips to the newly opened Pentagon.

In return, General Tire would become the majority shareholder. The principal shareholders were von Kármán, Zwicky, Malina, Parsons, Forman, Haley, and a handful of others. They argued over whether to accept the offer. Finally, they shrugged and said okay, at which point General Tire countered with a new offer, this time of only $50,000. The little management team at Aerojet was outraged. At the same time, they knew they were in a weak bargaining position. "We had exhausted our credit with the banks," von Kármán said, adding that "we would find it difficult to deliver on our contracts without outside financing."

After some additional back and forth, the Aerojet directors accepted $75,000 for 50 percent of the company's stock, which has to go down as one of the great business deals in that, or any, era. For General Tire, that is.

Even then, General Tire and its hard-driving founder, William O'Neill, who had been a Firestone tire franchisee before deciding to go into business against them, weren't satisfied. O'Neill wanted to get rid of the minority shareholders. Together, von Kármán's group owned 20 percent of the company's shares. O'Neil offered $350 per share. Now it was time for the Aerojet pioneers to be outraged all over again. "None of us wanted to sell," von Kármán said.

O'Neill threatened to merge Aerojet with other company units at an unfavorable stock price and to make other adjustments that would reduce the value of their shares far below $350. This didn't improve the attitude of a man like Zwicky, who hardly needed encouragement to dig in his heels over principle. There was talk of taking General Tire to court, but von Kármán wrote to Haley that his opinion about suing "was the same as the Chinese. You should not go to court in two cases—if you are right and if you are wrong. In all other cases it is permissible."[26]

That attitude sounds quaint today. But then, it seemed prudent to avoid a long court battle, especially in wartime; the Korean War was by then underway. Like the others, Zwicky agreed to sell his 125 shares, which netted him $50,000, some $400 per share. This was a very substantial sum for the times. But considering what Aerojet would become, there's no way to calculate his losses. Von Kármán did calculate, saying that if he had been able to hold onto his shares, they would have been worth $12 million. That was in the late sixties. In today's money, the value would be $118 million.

Starting with a work force of six, by the dawn of the 1960s, Aerojet General had grown to 41,000 employees, becoming the eighth largest company in California. In 1959, with the Cold War missile race under way, the company expected $300 million in annual sales. From the back room of a health food store where they sometimes had to hide from creditors, Aerojet occupied a 20,000-acre swath of land outside Sacramento. The Suicide Club became known as "the General Motors of the rocket industry."

"We have experienced failures and setbacks," Haley wrote in his 1944 report, as Aerojet stood on the brink of undreamed of success. But "the vista is very encouraging. The ultimate destiny of our work is second to no prospect in this amazing age of scientific achievement."[27]

Haley reflected a common belief that, despite the war being waged around the world, science would prevail over all obstacles. That the future was not nearly as bleak as it appeared. That when people like Fritz Zwicky could bring new things into the world that people had not imagined only a few years before, there was plenty of reason to go on believing that the future would be worth seeing.

A reporter visited Aerojet one day as it scrambled to meet its latest contract deadline. Wandering into Zwicky's office, he described a scene that captured the excitement of the new scientific age, when old problems would be solved and wonders created every day. "As the mild, tanned, dreamy-eyed professor puffs his old pipe from office to office at Aerojet's headquarters plant in Pasadena, he always has a piece of chalk in his pocket, and executives have learned that it is a good idea to have a blackboard handy when someone pops a question at Doctor Zwicky."[28]

The writer asked him how long it would take to get from New York to Los Angeles "when the ultimate in speed is achieved. . . . For about two minutes the professor chalked up equations soaring off into the stratosphere of relativity, where the ultimate vehicle would travel, of course."

His answer? Sixteen minutes. When the writer asked why it couldn't be fifteen minutes, Zwicky puffed and said, "If you made it fifteen minutes you burn up like a comet. That's about as fast as jets can go."

✳ 9 ✳

HOME FIRES

HIS SUCCESSES IN JET propulsion and rocketry gave Fritz Zwicky an entirely new status, bringing him an entirely new kind of celebrity. They also brought him contacts in military and government circles that would last the rest of his life, enabling him to play the kind of role in public affairs that he had always imagined for himself.

On the personal front, however, things were unraveling. His marriage, in particular, suffered from a kind of benign neglect that was all too common among astronomers. In Zwicky's case, the problem was doubled and tripled by the many hours he devoted to his other projects, preparing Pasadena for war and flying to and from Washington to lobby the government on his war plan.

In some ways, Dorothy and Fritz had made a good marriage, at least for the nine years it existed. Despite the differences in their backgrounds and personalities, they had a lot in common. They shared a love of travel and did a lot of it. Her diffidence was balanced by his passion. The more hyperbolic aspects of his personality rarely showed themselves when he was surrounded by family. This would be the case in his second marriage also. When he was confident of his footing, his natural charm and enthusiasm expressed itself in compassion and geniality. It was almost always in professional settings, when his vanity was pricked and his vision was dismissed, that he would erupt. And then, of course, *le déluge*.

Before the war, the couple hired an architect to design a house for them on land they bought in San Marino, now one of the elite communities outside Los Angeles. They planned to call it Castle in the Air; plans still exist, showing that it would have had all the modern angular dimensions and flow-through light features characteristic of the Frank Lloyd Wright designs that were so popular at the time. But after the war heated up, Zwicky didn't feel like going through with it. Shortly before their separation in 1941, the couple managed a last trip to Canada.[1] It didn't go well. The war was on everyone's mind. "People were so depressed everywhere that we were in the end glad to be home again," Zwicky wrote.[2]

Maybe there was no single reason for the rupture. Maybe it was just the standard set of disagreements and opinions, inflamed by the war. But in a note, Zwicky made a pointed reference to Dorothy's faith when he asked, "Where does the power of all quacks, Christian Scientists and 'healers' of all kinds come from?"[3]

The separation came as a rude shock to his family. "You can imagine what a surprise the contents (of the letter) were for me," Fridolin wrote in October 1941. His first thought was "what will grandmother say to this? We liked Dorothy very much from the first meeting on[;] she is someone you had to like, your grandmother too."[4]

Despite the split, and typical of Zwicky's fidelity toward those he drew into his life, he and Dorothy remained friends. He even took on her younger brother, Howard, as an assistant at Palomar. Despite Howard Gates's casual attitude toward work, Zwicky defended him and kept him on the payroll for more than a quarter century, until he himself was forced out.

Zwicky relocated to a "so-called Court which has six very nice large apartments,"[5] he wrote to Rösli Streiff. "I am the only man here, the others are all elderly ladies who are very concerned with my well-being."

One can only imagine: an eligible bachelor, earning a good living, famous and sought after, in their midst at a time when single men were being drafted by the thousands to aid the war effort. Once again alone, the Caltech professor was learning to take care of his own needs, using an iron and running the vacuum cleaner, at least when he wasn't roughing it at Palomar. "You probably can't imagine such hard work by a professor

in Switzerland, especially not by one who is involved in so many under-
takings as I have been these past years," he wrote in amused irony.[6]

He didn't much like cooking for himself, so he took some meals at the
Huntington Hotel, an elegant, Gilded Age edifice surrounded by twenty
acres of greenery and walking paths. Most of the time he ate at the Athenaeum, the Caltech dining club. Modeled after the ancient temple of
wisdom dedicated to Athena, the charming, adobe-style building was
built with a gift of $500,000 in stocks that had luckily been converted to
cash before the stock market crash of 1929. Albert Einstein was the guest
at the first formal dinner. Von Kármán was aware of Zwicky's rootless-
ness, and he made sure to include the younger man in social engagements,
which was why Zwicky was at the dinner table with Andy Haley for their
fateful meeting in 1942.

With gas and rubber being rationed, Zwicky bought a bicycle to ride
to the Institute for his classes. It was good exercise, he told friends. With
the war, his mountaineering expeditions had been interrupted, so the
bicycle at least gave him some needed physical exertion.

Despite his many other interests, he kept to his teaching schedule and
continued to make himself available to students. Even in his old age, he
was known among the students for having an open door policy. That
didn't mean Zwicky went easy on his charges. He was as demanding in
the classroom as he was in the lab.[7] His habit of mumbling in Swiss
German, as he scribbled equations on the blackboard, was well known
around campus. On one occasion, students brought a Swiss national to
class to listen to their teacher's rumblings. When they found out what he
was saying, they regretted their curiosity. "You stupid thickheads, you'll
never understand this anyway," was just one of his mutterings.[8]

With the war, however, there were fewer students. So many bright
young men were being drafted that student bodies at campuses all across
the nation were shrinking. One might think that Zwicky would be happy
to take a break from the burden of teaching, given his workload.

Certainly, Millikan thought so. In August of 1944, he sent Zwicky a
note informing him that since he was so much engaged in war work, the
university was granting him a furlough from teaching duties in the up-
coming school year. In a tart response, Zwicky said he had not requested
a leave and did not intend to take one.

Millikan replied that he thought the physics professor "would gladly follow" the common practice of other professors, who took leaves while they worked in the war effort. "None of the men who are employed on war projects at as much as four-thirds their peace-time salaries (Zwicky was earning more at Aerojet than in the classroom) remain on the payroll of the institute," he wrote.[9]

Zwicky said he was not in the full-time employment of any outside agency. "The 16 hours a week which I spend in consultation for pay is done in my spare time," he wrote. He worked much longer hours at Aerojet, but didn't admit that.[10]

After nineteen years supervising the stubborn Swiss, one would think Millikan would be less naïve, but apparently not. The dispute dragged on for weeks, growing ever more heated, until in late October, Millikan gave in. As far as he was concerned, Zwicky could remain on the payroll, in spite of the accepted practice.

Zwicky ran into trouble elsewhere as well. If he was a thorn in Millikan's side, he could be an even greater trial for the editors of scientific journals. His wide-ranging mind and the broad sweep of his theories were not made for the tight confines of journal reasoning. In 1940, as he shuttled to and from Washington on his autogyro campaign, he submitted a paper to the *Astrophysical Journal,* attempting to apply the principles of hydrodynamics to the motions of galaxies.[11] It was a stretch, and the referees who reviewed the theory were aghast. (Zwicky himself later referred to it as a "crazy paper."[12]) One urged outright rejection of the paper, criticizing the fact that the author had ignored "established methods" in favor of introducing new concepts "without an adequate or satisfactory analysis of all the implications" of these new ideas.[13]

The editor of the journal was the Russian-born astronomer Otto Struve, a man who could be every bit as intimidating as Zwicky.[14] He was also skeptical. But he soon gave way in the face of Zwicky's intransigence, even offering a kind of apology for challenging him. "I hope you understand," Struve wrote, "that this procedure of having articles refereed is a standard practice and implies no desire to criticize your work."[15]

One of the most bitter and long-running of his pitched battles erupted several years later over a paper Zwicky submitted to the Astronomical Society of the Pacific on the "morphological aspects of the distribution of clusters of galaxies." His argument that the tendency of galaxies to gather in clusters was common, rather than rare, would eventually be proven to be correct.[16]

The referees who reviewed the paper, however, were unconvinced and recommended against publication. They accused the author of, among other things, failing to explain what a cluster is, general vagueness, and drawing conclusions not supported by data. What followed was a two-year-long, increasingly angry exchange between Zwicky and the editor of the publication, Alexander Pogo. Before it was over, the dispute mushroomed to include Edwin Hubble; Ira S. Bowen, who had replaced Walter Adams as Mount Wilson's director; Vannevar Bush, the head of the Carnegie Institution in Washington, DC; and Lee DuBridge, who by this time had taken over from Millikan as president of Caltech. Pogo became so angry that, according to Zwicky's notes, he shouted into the telephone, "We are sick and tired of your paper!"[17]

For his part, Zwicky accused Pogo and Bowen—who had by then refused publication of several other Zwicky papers—of trying to suppress his work, adding, "your actions and those of Dr. Bowen with regard to the issue at hand, constitute a severe and dangerous attack on the freedom of scientific research and on academic and civilian freedom of speech in general."[18] That he was prepared to make the issue about freedom and censorship was very much in character. As he often did, when confronted with resistance, he invoked higher authorities, in this case, Albert Einstein. He said the great man had told him that his paper "represents one of the most ingenious, compact and concise approaches to cosmology."

Bowen, who was known both for his diplomacy and tenacity, it being said of him that the "difference between Bowen and a brick wall was that if you lean on the wall long enough, you can move it,"[19] warned Zwicky to back off. If he continued to write papers like that one, he would be laughed at, Bowen said.

If the Mount Wilson director thought that would cow Zwicky, he was mistaken. Zwicky wrote that "no first class researcher cares whether or not he is being laughed at." To Bowen's suggestion that he review the way other authors presented their ideas, Zwicky said that "much trash is being written by men who belong to what Einstein calls 'Mutual Admiration Societies,'" societies that Zwicky had no interest in joining.[20]

This conflict, in the end, caused the Mount Wilson and Palomar Observatories to rewrite their policies to allow staff to submit papers directly to the premier journals in the field without local review. It was a pattern that was repeated often in Zwicky's career. His willingness to fight every battle as blood sport was in keeping with his philosophy that accepted wisdom, whether handed down from so-called experts or, especially, superiors should be challenged. If it was a fight they wanted, it was a fight they'd get. Even if they didn't want it, as the peaceable Millikan clearly didn't, they were likely to get one anyhow.

Going along to get along was not Zwicky's style. He was in fact inordinately proud of this prickly tendency. He could recognize, of course, that he was alienating people, but he most often chose to see this as yet more proof of the envy of lesser minds. That he was allowed, throughout his life, to get away with this behavior was testament to his genius. In America, it has been proven that only the most gifted are allowed to behave in adulthood like adolescents.

But as they say about paranoiacs, sometimes they are right: They are after you. One famous case that Zwicky referred to over the years concerned a paper of Hubble's,[21] which dealt with what came to be known as dwarf galaxies. Hubble suggested that these small systems might play a "significant role" in cosmological theories. He was right.

Zwicky had been studying those groupings for some time and recognized very early that they might be important. Yet in his paper, Hubble credited Baade for the discovery of two "unusually interesting" systems, which he labeled, "Baade's System in Sextans" and "Baade's System in Leo." In fact, both had been discovered by Zwicky, using the 18-inch Schmidt telescope, and only confirmed by Baade with the 100-inch Mount Wilson instrument. Zwicky was convinced, also correctly, that these small systems were important to understanding galaxy formation as a whole.[22]

That Hubble attributed Zwicky's work to Baade may simply have been an oversight, as he claimed. If so, it was a pretty serious one. Or it may have been evidence of Zwicky's increasing alienation from his colleagues in astrophysics. Hubble humbly asked *Scientific Monthly* to print a correction in 1941.

"This, however, in my career as a physicist and astronomer is one of the comparatively rare incidents in the USA in which the gentlemanly spirit upheld by so many of our great predecessors . . . prevailed," Zwicky wrote many years later.[23]

Busy as he was, and with the war disrupting the mails, communications between Zwicky and his family in Europe were sporadic. While Fritz was in Washington in 1944, his father managed to get a letter out for the first time in months. He told his son that he had been forced to return to Glarus; Zwicky's brother Rudi remained in Bulgaria, trying to keep the business going. But that had become a dangerous place, with Allied bombs landing in the capital, Sofia, and the Soviet army, fresh from its victory in Stalingrad, approaching from the east.

There was plenty of reason to feel anxious about the future, but Fridolin Zwicky was a hopeful man. Perhaps things would return to normal soon. For the time being, however, their business affairs in Bulgaria must await the end of "this terrible war."[24]

Near the end of his letter, Fridolin wondered about his son's future plans. "You haven't given up your dream, have you, of returning to Switzerland?" This was an old topic. Zwicky had barely arrived in America when Fridolin began urging him to return. Now that his father was getting on in years—he would be 76 in August—his appeals became even more frequent. In reply, Fritz cautioned against counting on him to come home any time soon. Travel to Europe during the war was "extremely difficult," he had written.[25] He was also occupied more than full time at Aerojet. As for landing a teaching job in Switzerland, he wasn't sure how his experience and knowledge would transfer. The truth was that Fritz Zwicky, despite his longings, only seriously entertained returning to Europe when he was feeling low. Most of the time, he was well aware of how lucky he was to be where he was. With access to the two big telescopes on Mount Wilson and the 18-inch Schmidt at Palomar, he was at the center of celestial exploration on Earth.

Even more important, work was well under way on the great telescope on Palomar Mountain, the 200-inch Big Eye. At the time of his writing, the giant, twenty-ton mirror blank rested in Caltech's optical shop. Any time he wanted, Zwicky could visit the shop and see the great chunk of glass, the grinding of which had been temporarily halted by the war. When it was finished, the parabolic surface would be accurate to two-millionths of an inch. That was a tenth the size of a wavelength of visible light (500 angstroms, in the language of astronomy). Zwicky knew better than most what the new telescope would do. He had posed some of the deepest questions that it would answer. The idea that he would leave all that behind was in the realm of fantasy, even though it was not something he could explain to his father.

In closing his letter, Fridolin said he was "in excellent health" and hoped the same was true of his son.

That would change before the first Soviet boots tramped into Sofia in September 1944. During the spring, Fridolin developed a strange skin disease. Blisters appeared on his wrist and neck. Fritz's younger brother, Rudi, wrote with updates on Fridolin's condition. He wasn't in great pain, but when the blisters grew and spread, Rudi had him hospitalized.

After a few weeks, the problem appeared to clear up, and Fridolin was allowed to get up and move around. But the hope of a cure vanished when the blisters appeared again, this time spreading all over his body. By now, Rudi was so worried that he put business aside and moved to Glarus to be near his father.

Fridolin suffered from pemphigus, an autoimmune disease in which the body's own antibodies attack the "glue" binding the epidermal cells together. Basically, the body tries to slough off its own skin. The treatment for pemphigus is the administration of steroids, such as cortisone. At the time, this treatment was unknown. The great irony, which Zwicky noted later, was that his close friend Tadeusz Reichstein would be the one to find a way to isolate and replicate cortisone. "Is this not a peculiar tragedy of circumstances?" Fritz asked, in a letter to his brother.[26]

On July 7, 1944, Rudi wrote, Fridolin "was good. He ordered a bottle of wine" to drink with his friends and family the next day. At midnight, Rudi got a call that Fridolin was failing. By 1 A.M. on July 8, he was "asleep forever." The doctor assured Rudi it was for the best, because this disease can make one utterly miserable. Fridolin Zwicky was cremated and his ashes interred with their mother, Franziska.

Rudi was confident their father's care was good, but he bemoaned the fact that Fridolin never realized "his wish to return to Varna."

Many people—Swiss dignitaries, business people, and politicians who recalled Fridolin Zwicky's diplomatic work—attended the funeral. It was left to him, Rudi wrote, to administer the estate according to their father's wishes and to try to keep the business going. Not having his father's sunny disposition, Rudi was not optimistic. He said it was "unthinkable"[27] to live under Communist rule. He would not, therefore, return to Varna.

Death did more than remove the foundation of the Zwicky family. It exposed fractures that only their father's presence had kept hidden. Without Fridolin's guidance, Rudi and his sister, Leonie, fell to squabbling over their father's meager assets. Fritz sent word that he would gladly give up any claim to the estate if that would help things, and he warned them against involving the courts in the mess. The arrival of Communism made it unlikely that they would find justice there.

"If all property is being eliminated, you will not get anything anyway and fighting for it with Rudi is useless," he wrote to Leonie. "For me personally, both yours and Rudi's interest in money matters seems somewhat exaggerated."[28]

This was perhaps easier for Fritz to say than for his siblings to hear. His income from Caltech and Aerojet made it difficult for him to appreciate the full extent of the hardship his relatives were facing in a war-torn region living under Communism. In 1947 alone, Fritz earned $20,800—no fortune, but enough to make him quite comfortable at a time when the average annual household income in America was $3,100. Rudi pointed out that Fritz had a pension and savings to protect him in his old age. He and Leonie had none of that.

When Soviet troops arrived in Bulgaria, they were at first welcomed as liberators. Bulgaria held the earliest war crimes trials in Europe, executing almost 3,000 German collaborators. But because the country emerged from the war with no political structure, the Communists seized their opportunity in 1946, installing one-party rule that lasted until the Soviet Union began to collapse.

While Rudi Zwicky remained in Zurich, Leonie stayed with her husband, Stefan Staneff, in Varna, at the family home. The small but sturdy stone house was a charming structure surrounded by gardens and bore a crest over the doorway representing Fridolin's consular past. It eventually proved too much of a temptation for the new Communist rulers, who evicted them. The lead shot factory was also seized.

Personal hardships multiplied. There were food shortages, and common household needs became impossible to get. Fritz sent packages of food and other supplies, but it was hard to know what, if anything, got through.

In his letters, Rudi railed against the Americans for allowing the Communists to take over Eastern Europe. "The Russians will not rest until Communism spreads around the world," wrote his wife and Fritz's sister-in-law, Elfriede.[29] Rudi suffered from heart trouble and worried what would happen to Elfriede when he was gone. His flight from the Communists had cost him his livelihood, so he was starting over in Zurich, "almost from scratch because in Varna it will not be better soon." He hoped to start a business in nylon stockings, which were particularly hard to get once the Iron Curtain descended. Fritz connected him with a supplier.

The feud between the siblings in Europe dragged on for years. Fritz made several peace-making efforts, even turning over his claim to the family vineyard in Tschair, near Varna, to his sister.[30] But that did little to smooth relationships that had been fractured by war, privation, the failed promise of the socialist paradise, and simple family antagonisms.

In this atmosphere of familial tension, the brothers exchanged views on the survival of the Zwicky name, in danger after Fritz's divorce. "As you say, the Zwicky line dies with us," Rudi, who was childless, wrote at the end of the war. "You need to hurry because in three days you are already 47 years old."[31]

Something would soon be done to address that problem.

* 10 *

SECRET MISSIONS, FINDING VON BRAUN

IN THE LATE SUMMER of 1944, Army Air Force General Henry H. "Hap" Arnold called von Kármán to a secret meeting, held on an empty runway at New York's La Guardia Airport. "We have won this war and I am no longer interested in it," Arnold said bluntly. The D-Day invasion had taken place only three months earlier, and much fighting remained, but Arnold was already looking ahead.

"I do not think we should spend time debating whether we obtained the victory by sheer power or by some qualitative superiority," he told von Kármán. "Only one thing should concern us. What is the future of air power and aerial warfare? What is the bearing of the new inventions, such as jet propulsion, rockets, radar, and the other electronic devices?"

Von Kármán was floored. "I had always admired Arnold's great vision, but I think then that I was more impressed than ever," he wrote in his autobiography.[1]

General Arnold was a true aviation pioneer, with the vision to match. The son of a Pennsylvania physician, Arnold received flight instruction at the Wright Brothers aviation school in Ohio, becoming one of the world's first military pilots. Under his leadership, the United States would develop the intercontinental bomber and the jet fighter and make improvements in radar and atomic warfare. With a character too broadly inventive to be confined to warmaking, he wrote children's books on the side.

His purpose in meeting with von Kármán was to get scientists in America thinking about the future.

"I see a manless air force," Arnold told a group of scientists whom von Kármán assembled in Washington, DC, a few weeks later. "I see no excuse for men in fighter planes to shoot down bombers. When you lose a bomber, it is a loss of seven thousand to forty thousand man hours, but this crazy thing (the German V-2 rockets, which had begun landing on London) they shoot over there takes only a thousand man hours (to build). For twenty years, the air force has been built around pilots, pilots and more pilots. . . . The next air force is going to be built around scientists."

When Hitler's thousand-year reich finally collapsed in May 1945, Arnold went to von Kármán with a new proposal: "Why not go to Germany and find out first-hand how far the Germans actually have gotten in research and development?"

Von Kármán thought Arnold's idea was a "capital" one. Some last-ditch fighting was still going on in Germany when von Kármán flew with several associates to Paris on a military C-54. Separately, on von Kármán's suggestion, Arnold asked Zwicky to make the same trip, but with a different team.[2] It was an inspired choice, because at that time, Fritz Zwicky knew more about rocket and jet propulsion than almost anyone else in the United States. It was also a controversial one, and not only because of Zwicky's tendency to go his own way. Unlike many other members of the European intelligentsia who emigrated to America, Zwicky chose not to become a citizen of his adopted country, something that would cause him great trouble during the McCarthy era a decade later. The fact that Arnold overlooked that problem was a testament both to his trust in the Swiss professor and to his considerable influence in military circles.

Zwicky and von Kármán traveled under cover, disguised as military officers. Von Kármán bore the rank of major general, while Zwicky's card for the Consolidated Officer's Mess was stamped "Colonel's Privilege."

On arriving in Paris, Arnold's teams of scientists found that the Peenemünde rocket complex, located on an island in the Baltic, was under Soviet occupation. This was a potential disaster. If the German research, and, just as important, German researchers, fell into Soviet hands, it was feared that the world would soon face a threat nearly as grave as the Nazis.

Luckily for the Americans, most of the important German engineers had been evacuated from Peenemünde and were being detained at Garmisch-Partenkirchen, a ski resort several hundred miles to the south. On May 9, 1945, two days after Germany surrendered, Zwicky and his team traveled to the 6th United States Army Group's Intelligence Specialists Camp, and from there to Garmisch. There he began weeks of debriefings with the cream of the German rocket corps, the results of which were included in Zwicky's classified 188-page analysis, *Report on Certain Phases of War Research in Germany.*[3] Among those he interrogated were General Walter Dornberger, the man in charge at Peenemünde, and Wernher von Braun, the tall, socially adept chief scientist at the rocket complex. Von Braun would play a crucial role in the development of the American space program after the war.

In a 1951 interview with a writer for the *New Yorker,*[4] his first extensive conversation with an American reporter, von Braun described the German scientists' flight by railroad from Peenemünde. "Russians were only a hundred miles away, and we could already see an Iron Curtain was coming down," he said. He and Dornberger "wanted our outfit to fall into American hands," rather than the Russians. So they and several hundred of their lieutenants headed for Bavaria, to await American troops. "There I was, living royally in a ski hotel on a mountain plateau," von Braun said brightly.

The Peenemünde crowd learned over the radio of Hitler's suicide, that the war was over and an armistice had been signed. Still there was no sign of the US Army, so von Braun sent his brother down the mountain on a bicycle to search for the Americans. After some confusion, von Braun and his associates were put into the hands of Operation Paperclip, the secret organization given the task of deciding which German scientists would be repatriated in America, just hours before the arrival of Soviet forces. The first to question von Braun, the first to realize the great prize they had, were Fritz Zwicky and two associates, Richard Porter of General Electric and Clark Millikan, the son of Robert Millikan, who was by then teaching at Caltech.

"Their questioning, of course, was extremely intelligent," von Braun said. "These men are top scientists. I (will) do business with them."[5] Do business he did, rising so fast and so far that when Explorer 1 was launched

Zwicky in uniform at Garmisch-Partenkirchen, the ski resort where the top German engineers behind the fearsome V-2 rockets were being held at the end of World War II. Zwicky interrogated hundreds of them, including Wernher von Braun, to produce his influential report on Nazi science. Standing next to him is Colonel John O'Mara, the man in charge at Garmisch, who would become a close friend and fellow anti-Communist Cold Warrior.

in 1958, von Braun, the ex-Nazi who built the revenge weapon that reduced much of London to rubble, was allowed to pose prominently for the cameras. Smiling broadly, he held aloft a model of the spacecraft, proof that America was finally, after the Sputnik disaster, ready to compete in the space race with the Soviet Union.

From late April until the end of July 1945, Fritz Zwicky interrogated hundreds of German engineers, toured the rocket facilities at Peenemünde, studied the giant wind tunnels in southern Germany, and ordered them disassembled and shipped back to the United States. They would be used to design the next generation of American fighters and bombers. He questioned test pilots and uncovered plans for advanced weapons that never reached the manufacturing stage, such as a long-range missile—the world's first intercontinental ballistic missile (ICBM)—that could reach New York. Trajectories had already been computed for such a terror weapon when the war ended.

Zwicky was deeply impressed by the scale and scope of the German achievements, noting in his report that the Germans were ten years ahead of the United States in rocket research. Given his work at Aerojet, he was well equipped to know just how far behind the Americans were. He found shortcomings, as well, however, writing that there was no evidence the Germans ever considered replacing carbon in their fuels with light metals—something Zwicky did at Aerojet—to achieve greater thrust. The V-2 utilized methanol and ethanol as fuels in combination with liquid oxygen as the oxidizer.

The success of the V-2 program, he wrote, was not the result of an inspired genius, no matter what the American press was saying about the charming and voluble von Braun, the wunderkind of German science. It was achieved through the "enthusiastic cooperation of a large number of only moderately competent . . . individuals." Even so, he said, once he had seen what the Germans had accomplished, he found himself "hard put for an explanation of why the enemy did lose the war."[6]

There was no lack of inventiveness, nor was there a shortage of funding. According to Dornberger and von Braun, the German engineers got whatever they wanted. Von Braun had been recruited out of the University of Berlin in 1932 and given free rein to develop his rockets. He claimed his interest was in puncturing space, but Hitler had other ideas. The German scientists were even relieved of the burden of writing weekly and monthly reports, something Zwicky had to do in America. "From my investigations of the German and (later) Japanese scientific and technical potential it became clear to me that America has no monopoly on brains," he wrote later.[7]

In America, he said, the chief danger to "irregulars" like him, who proposed disruptive policies and theories, is that they would be "considered a charlatan or wild man."

Bad as that was, the danger was much greater in a dictatorship like Hitler's Germany and Stalin's Soviet Union. Because the penalty for failure might well be death, a scientist might hesitate to put forward new ideas. Zwicky found several examples where the Germans failed to capitalize on potential breakthroughs, such as jet fighters, supersonic guided missiles, and fin-stabilized artillery projectiles. The greatest example, however, was the atomic bomb. As early as 1938, the German chemist and physicist, Otto Hahn, discovered nuclear fission.[8] His 1939 paper on its potential influenced Albert Einstein, who went to President Roosevelt and proposed that America get to work on an A-bomb before anyone else did. Roosevelt listened and acted, setting up the Manhattan Project; in contrast, Hitler flew into a rage. "Why waste money? Give me jets. Give me mines. By the time you work out this atomic energy, the world will be at my feet."[9]

Albert Speer, Hitler's minister of armaments, said the Führer had a "fundamental distrust of all innovations which, as in the case of jet aircraft or atom bombs," were outside his experience in World War I.[10]

"In spite of all other mistakes made," Zwicky concluded in his report, "the Nazis might well have won the war if the German scientists and engineers had dared to literally take their lives into their hands and force into the production line and into the field of combat these inventions which they had developed."[11]

Zwicky's lengthy, detailed report was widely circulated among the upper echelon of Army Air Force offices, prompting a note of lavish thanks from Arnold himself. It was instrumental, along with Zwicky's later report on the destruction caused by the A-bombs dropped on Japan, in President Harry S. Truman's decision to award him the Medal of Freedom in 1949. The highest civilian honor, Fritz Zwicky was the first foreigner to receive it.

Zwicky was impressed by the cooperation he received from the Germans, who seemed almost eager to describe their achievements, like students hoping for a good grade. He was irritated, on the other hand, by the hamhandedness of the American troops, who destroyed much of the equipment they found in the German laboratories. Von Kármán was appalled by the German attitudes. Hearing that American troops had discovered a completely unknown institute in a pine forest near Braunschweig, in northern Germany, von Kármán hurried there. While bouncing along a rutted road in a jeep on the way to Volkenrode, he found himself depressed by the devastation he saw all around him, caused by the Nazi "urge to war."

"I was deep in a country which I had believed was the fulcrum of world science," he wrote in his autobiography. "Now it was rubble, abandoned machinery and dull spiritless faces along the road." As a V formation of American planes flew overhead, final proof of victory, a group of Germans crowded around his jeep "and to my amazement sought to shake hands. It was as if we had just won a tennis match. I was shocked. I rose in my seat and I am afraid I gave way to feeling. I lectured them on remorse, and on humility in the face of what they had caused. Where was the conscience of the German people?"[12]

After reviewing the secret installation at Volkenrode, von Kármán came to believe, like Zwicky, that the Germans could have won the war. The fifty-six buildings he toured housed research facilities in ballistics, aerodynamics, and new types of engines. Also like Zwicky, von Kármán came to believe that while the research was good, it was not spectacular, and "largely the result of lavish funding rather than superior ability."[13]

In a dry well—the Germans had been ordered by the SS, Hitler's elite paramilitary force, to destroy their research, but some was hidden, preserved in the hope Nazism would rise again—von Kármán found wind tunnel data for a new type of airplane with swept-back wings. He knew aerodynamicists in America had considered just such a configuration, but the work was in the nascent stage. In Germany, they had already tested the design and completed plans to build a supersonic jet fighter. This discovery led to the development of the B-47, the first US swept-back bomber.[14]

After finishing his work at Volkenrode, von Kármán traveled fifty miles south to Nordhausen, in the legendary Harz mountains, steeped in folklore and myth, where Goethe's Faust was damned on Walpurgis Night. The experience was a "ghastly" one for von Kármán. Here, in a labyrinth of underground salt mines, a thousand slave laborers built the V-2s. He discovered that the Nazis in charge at the mines had developed a plan for starving the workers to death on a minutely constructed schedule. Each worker was given so many months to live, and his food ration was steadily decreased so that he died exactly on schedule. It was "a perversion of science beyond anyone's nightmarish imagination," he wrote.[15] Of course, he was unaware of even worse conditions that prevailed in the concentration camps, where millions of Jews, homosexuals, and other "enemies of the state" were exterminated with the same barbaric efficiency.

In September 1945, when von Kármán, Zwicky, and others from the institute had returned to Pasadena, Robert Millikan scheduled an evening talk before the Caltech community. "Long before the collapse of the axis in Europe, plans had been made by both the navy and the army to send more than a dozen members of the institute staff into Germany and the occupied countries to gather information as to conditions existing there, particularly with reference to scientific and engineering developments . . . some members of this group were among the very first to follow the American push into central Europe," Millikan wrote.[16] Scheduled for the session were Zwicky; Millikan's son, Clark; and H. D. Tsien, another early rocket pioneer at Caltech. Tsien would eventually emigrate to China, where he helped guide that nation's rocket research, earning the sobriquet "Father of Chinese Rocketry."

But Zwicky's service in the war effort was not yet complete. When he discovered the buzzbomb facilities in Germany, he knew immediately what this weapon was. He had pitched the idea to the military in Washington. Realizing this weapon could be put to good use in the still-raging war with Japan, he suggested that the V-1s be sent back to the United States with all dispatch. Despite his contacts in the defense establishment, he appeared not to know that the United States was in the final stages of completing an even more powerful weapon. The military officers with whom he worked in the Navy and Army Air Force were also either in the dark or unsure that the atomic bombs—Fat Man and Little Boy—would be ready in time. Letters in the Zwicky archive attest to their anxiety that his efforts be given the highest priority, so that the "intermittent motors," as they were called, could be used to attack Japan from offshore.

Because Zwicky's researches were "of the utmost importance to the U.S. Navy in the prosecution of the war against Japan," wrote the United States Naval Technical Mission in Europe,[17] he should be given complete access, particularly in the Munich manufacturing zone, to whatever installations he required.

As it turned out, the American version of the buzzbomb was not needed. After the atomic bombs fell, the Japanese empire collapsed. Zwicky was now chosen for another fact-finding trip, this one to the bombed-out cities of Hiroshima and Nagasaki. Originally, von Kármán planned to go as well, but after the 59-year-old General Arnold suffered his fourth heart attack during the war, the military thought it prudent for von Kármán to remain in the United States to write a report of his findings. Arnold's uniquely far-reaching vision—among other things, he helped found Pan American World Airways and Project RAND, which became the admired think-tank Rand Corporation—was considered so important to postwar planning that it was considered critical for von Kármán to put his ideas and suggestions in writing before the general died.

Von Kármán's report, *Toward New Horizons,* would become one of the seminal documents in the history of American military preparedness. It provided the guideposts that would enable the United States to dominate the postwar world in air power. The introductory volume of the report would be simply titled *Science—The Key to Air Supremacy.*

While von Kármán began to write, Zwicky flew to Japan with two colleagues from Caltech,[18] arriving three months after the bombs were dropped in early August 1945. He was among the first Western intellectuals to see the devastation firsthand and to take note of the vast numbers of victims. Notes he preserved from that trip give an accounting of the loss of life and structural damage to the dwellings. The numbers were staggering, yet, typical of his analytic mind, he noted them without comment. Under the heading, Hiroshima, he wrote, "Houses, Burnt, 55,000; Half-burnt, 12,600." Altogether, he recorded 78,170 structures damaged and destroyed, in a city of around 255,000. Under the heading, "Persons," he wrote 46,185 killed, 21,125 males, 21,277 females, and 3,773 "uncertain."[19] Meaning, of course, that their bodies were so atomized that their sex could not be determined.

Besides touring the blast sites—he returned with several souvenirs from a decimated temple near Ground Zero, charred bricks and a finial— Zwicky conferred with Japanese nuclear scientists at Kyoto University. He discovered they, too, were working on a nuclear device.

What he saw in Japan turned him into a lifelong critic of the decision to drop the bombs and of all those, such as Oppenheimer, head of the

An unidentified man inspects the destruction near Ground Zero in Hiroshima, three months after the atomic bombs were dropped on Japan, in August 1945. The devastation Zwicky saw on this fact-finding trip for the US government made him a lifelong critic of the decision to drop the bombs.

Los Alamos, New Mexico, laboratory where the bomb was designed, who recommended its use.

"The decision to use the bomb as it was used may . . . turn out to be one of the most disastrous ever made,"[20] Zwicky wrote in his 1957 book, *Morphological Astronomy*. He took special note of the irony that Oppenheimer and Arthur H. Compton, who led the Manhattan Project's Metallurgical Laboratory, had been among the leaders in the American Association of Scientific Workers, the group behind the 1940 Peace Resolution aimed at keeping America out of the war. Having seen the devastation up close, only weeks after the bombs were dropped, he could never be convinced that it was the right thing to do, even if it did save the lives of thousands of American servicemen who would have had to fight their way onto the Japanese homeland. For him, dropping the bombs cost the United States the moral high ground in international affairs, and that was too high a price to pay.

In their defense, Compton and Oppenheimer denied arriving at their decision with anything like the flippancy Zwicky alleged. Besides preventing the deaths of American service personnel, the "tempered" use of a modest atomic weapon, they said, would, in the end, save Japanese lives, since Emperor Hirohito would undoubtedly have demanded the ultimate sacrifice from his people to preserve their sacred land. Zwicky found those arguments unconvincing.

"It is one thing to decide to use a new and horrible weapon," Zwicky said, "and quite another to be reluctant to develop such weapons so that free men shall be prepared to meet whatever shall be brought against them."[21] In his mind, such weapons should be reserved only for defensive use.

Zwicky's problems with Oppenheimer went beyond his role in the use of the atomic bomb. Noting Oppenheimer's socialist views, he referred to him as a "fellow traveler," a smear popular at the time that was used against those who were thought to be sympathetic to Communism. For Zwicky, whose family lost everything when the Communists took over in Bulgaria, there was hardly a worse epithet.

Oppenheimer willingly applied the term to himself, defining it as someone who is sympathetic to Communist philosophy but not willing to blindly follow dictates from Moscow. His opinion was that Americans were unrealistically fearful of Communism. Invited to speak at Caltech during the chilliest period of the Cold War, he told the story of a terrified woman who wakes from a dream to see a bloodthirsty demon at the foot of her bed. "My God," she screamed, "what are you going to do to me?" "Madam," the demon replied, "this is your dream."[22]

For Oppenheimer, it was America's fear that turned the Soviets into the great enemy. Zwicky, who had watched Lenin close up and was well aware of Stalin's grotesqueries, considered Oppenheimer dangerously naïve.

One of their most public clashes came over the question of how long it would take the Soviets to get "the Bomb." Oppenheimer, trying to soothe American fears, said it would be many years off. He "nearly jumped down my throat," Zwicky said, when the Swiss argued that the Soviets could probably develop an atomic bomb in three years.[23] Oppenheimer's opinion was by far the more influential, since it fit with America's prevailing attitude that the Russians were brave but stupid. In the end, Zwicky was proved right when the Soviet Union conducted its first nuclear test in 1949, four years after the war ended.

Unlike Zwicky's report on German war research, his report about the effects of the bombing of Nagasaki and Hiroshima was not declassified, in part because it came to the unthinkable conclusion that it was possible to prepare for, and survive, atomic attack. In the aftermath of World War II, the world was seized by revulsion over the devastation caused by the Bomb. Schoolchildren in America were taught to "duck and cover" under their desks, and an entire generation went to bed dreaming of mushroom clouds. To counter fears of Armageddon, military and government officials launched a campaign to calm the waters. Zwicky belonged to this group, who felt that, terrible as the weapon was, it was survivable, given proper preparations. He had seen how quickly the Japanese rebuilt.

Although Zwicky's A-bomb report remained secret, Major Alexander de Seversky took up the theme in an infamous article in the February 1946 issue of *Reader's Digest*. In it, he lamented the "near hysteria at the first exhibits of atomic destruction."[24] Hiroshima suffered because of bad construction, he said, likening the bomb to "a great fly swatter, two miles broad, (which) slapped down on a city of flimsy, half-rotted wooden houses and rickety buildings." If such a bomb landed on an American city, Seversky argued, the damage would have been "limited to broken glass over a wide area."

To children ducking and covering, this might have been welcome news, but Seversky was quickly shouted down, his conclusions roundly attacked and mocked. Within a few months, the magazine was forced to print a much more accurate, and much more frightening, assessment of what a nuclear bomb would do to an American city.

Zwicky remained a believer, however, that the danger from atomic bombs must not be overestimated. Aerojet, which soon began building ICBMs with nuclear payloads that targeted a sixth of the world's surface, jumped into the debate, publishing a how-to pamphlet.[25] It was of the "our friend the atom" variety. Accompanied by drawings showing a family rushing into each other's arms, it offered helpful tips, recommending that when the yellow alert sounded, people should tune the radio to CONELRAD, the emergency alert broadcasting system, stock the car with food rations—though it said nothing about where they should go—and find the best inside shelter. When the three-minute "take cover" signal sounded, they must lie on the floor and stay put until "the blast subsided." The last bit of advice: "Remain calm."

There was no need to build an expensive bomb shelter, which many did in the fifties. According to the writer Walter F. Libby, spending as little as thirty dollars, you could construct your own backyard shelter with sandbags.[26]

In a letter to von Kármán after the *Reader's Digest* article, Zwicky noted[27] that Seversky "has stolen part of our thunder as to the real effects of the atomic bombs on Hiroshima and Nagasaki. Although there are some serious errors in his article, its publication is probably a good thing since it will help to counteract the defeatist attitude which even scientists seem to have that there is no effective defense against atom bombs."

Zwicky's pragmatic nature influenced his attitude. He was convinced that, having been used once, the Bomb would surely be used again, this time in all probability against the United States itself. If it was to be employed, it was important to get used to the idea, to prepare for it. Others argued that the sheer horror of the bombs dropped on Japan would serve as the best warning against their ever being used again. To Zwicky, that attitude was folly. In the decades since, the policy of mutual assured destruction has succeeded in preventing a second use of nuclear weapons. For how long? Zwicky would ask, were he around. He never understood why his advice about defending against nuclear attack was not taken. Years later, he railed to friends that his suggestions had "gone unheeded, even to this day."

In a more philosophic mood, if not a more optimistic one, he jotted down in one of his many diaries his thoughts about the pickle mankind had got itself into with the Bomb. "It is a simple observation that throughout the history of mankind, it proved to be impossible to have a decisive majority of men agree on anything. The present generation finally seems to have achieved this goal. The agreement reached, however, is of a peculiar kind. It consists of the admission that all of us, individually and collectively, find ourselves in a general mess."[28]

For the public, the atomic bombs dropped on Japan showed that the new "scientific age" might not, after all, solve all problems. It might, in fact, bring an entirely new and even more dangerous set of difficulties. Some scientists, especially the younger generation, began pushing back against the idea that science would bring a utopian age of ease and plenty. They also began questioning the celebrity worship lavished on scientists such as Einstein and, to a lesser degree, Zwicky.

Richard Feynman, at the time a young colleague at Caltech, said something on the subject that infuriated Zwicky, which the latter never forgot. When it came to politics and social issues in general, he warned, the scientist was "as dumb as the next guy."[29] Feynman, a Manhattan Project scientist, cautioned against expecting people like him to lead the way to a better future.

Zwicky responded angrily. The "scientist cannot and should not avoid playing an important part in non-scientific affairs," he wrote. To Zwicky, who had leapt in to solve problems far outside his field of expertise, Feynman was abdicating his responsibilities as an educated and respected member of society. As such, he had a special duty to raise his voice and help his less-favored fellow citizens to see the light of reason and good policy.

"Expressing views of this sort to some of my colleagues during the war, I was challenged to demonstrate," Zwicky wrote. Which is what he did when he tried to influence defense policy before the war. And this view governed his confidence in his ability to enter a new field—jet propulsion, for which he had not prepared—and make fundamental breakthroughs. His accomplishments showed "that a scientist can venture beyond the borders of his own field without disastrous results, but with useful results," Zwicky wrote.[30]

"There is no reason for us to act like spiritual cowards, pretending that we are as dumb as the next guy."[31] He could not, and would not, believe he was as dumb as the next guy. About anything.

While he was in Europe in the fall of 1945, Zwicky made time to pay a visit to his friend Rösli Streiff. In a subsequent letter, he mentioned a "pleasant last evening in Zurich." Considering himself a gentleman of the old school, he would not state bluntly details of what transpired on that pleasant evening, even to Streiff herself. But he did note with satisfaction that now that he was "no longer married I am a free man again and can resume friendships with all my old and new friends."[32]

He was exhausted by all the travel, and by the sorrow of seeing the destruction of warfare firsthand. Writing from a hotel room in Washington, DC, he said he had suffered many sleepless nights and had lost weight. Still ahead was writing up his findings. When he was done, he hoped "to go back to my own ideas for a few months," while recovering in his Pasadena apartment, under the gentle care of his matronly neighbors.

But he was also doubtful he could ever fully return to the simple confines of the academic life he had known before, because "the world situation does not let one rest. . . . If you see the causes of tragedies so clearly, and still have a spark of energy you can't sit quietly by."

With all these things going on, Zwicky's observing time at Palomar was limited. On one occasion, he escorted Charles Galton Darwin, grandson of the man who changed for all time humanity's ideas about its place in the animal kingdom, to the site where the great 200-inch Hale telescope was rising from the forest. This Darwin was himself a notable member of the scientific establishment, winning the Royal Medal of the Royal Society for his contributions to quantum mechanics. "In my enthusiasm," Zwicky wrote of the visit, "I claimed that you could discover a new cosmic object every day." Darwin replied drolly, "Why don't you then?"[33]

Despite his achievements during the war, Zwicky found himself in a melancholy mood at the dawn of 1946. He fumed that the advice he had given about defensive preparations for nuclear attack had been "sneered at" by the leadership in America. He recalled being ridiculed for suggesting in 1940 that the Allies could win the war with five divisions and 100,000 men, if they only took his advice about employing super-gyros. In a self-pitying note to Streiff, he wrote that "mountains were my life, but that's all in the past."

Climbing, he said, would "probably finish off my heart and kidneys." The note was partly designed to respond to Streiff's desire to see him more often, but his sour mood mushroomed into a harsh critique of his life and character.

"I don't have any concrete plans for further ahead. . . . So please don't ask . . . I can't force solutions, and when I have tried in the past, it has always gone wrong. I don't know whether I will be doing astronomy or nuclear propulsion for rockets, or whether I will be in Switzerland or China, with which people I'll be living in a parallel existence, just like the two sailing boats in C. F. Meyer's wonderful poem. Maybe the answer lies in the stars."[34]

Indeed.

* 11 *

THE MARCH INTO SPACE

ZWICKY SOON RECOVERED from his melancholy. There was work to be done, for the war never really ended for him. In early 1946, Colonel John ("Jack") O'Mara, the commanding officer at the Garmisch compound where Zwicky questioned the German V-2 scientists, sent him a letter asking for his help. The fondness between the two men, developed in Germany, would only deepen as the years went on. O'Mara not only admired his friend's imagination and technical skill, but he also bought wholly into Zwicky's Renaissance man persona. He even adopted Zwicky's romantic terminology, referring to his friend as a lonely "free agent" in a gray flannel world beset by hesitation and weakness.

In his letter, O'Mara asked Zwicky to recommend "a few qualified people that we might contact and perhaps snare into the position of running" a guided missile department at Wright Field.[1] The German V-1 buzzbombs that Zwicky had shipped to the United States at the end of the war were being stored and tested at the Ohio base.

O'Mara's inquiry was part of a struggle that was already breaking out in the Pentagon over which service would control the development of the nation's trillion-dollar missile fleet. Over the next two decades, the United States would deploy 30,000 nuclear warheads in ICBM silos across the Great Plains, on Polaris submarines, and lashed into B-52 Stratofortress attack bombers. The little rockets von Kármán's Suicide Club toyed with in the drywash outside Pasadena would mutate into the most fearsome arsenal ever devised by humankind. Together, these missiles would be capable of extinguishing life on Earth many times over. Aerojet, the rocket company started in a Pasadena storefront, would earn hundreds of millions of dollars building parts of those missiles.

Who or what Zwicky recommended to his friend is not known, but O'Mara eventually moved on to a new duty, supervising Project Blue Book, the Air Force investigation into UFO sightings. Once there, he turned to his old friend once more, this time to help him formulate a policy on the so-called alien visitations being reported across the nation in the wake of the famous 1947 "crash" at Roswell, New Mexico.[2] In recent years, ufologists have claimed that O'Mara privately believed that UFOs were real. It's clear from his and Zwicky's correspondence that he did not. But he wanted to say something more to those who were sure they had seen alien craft than "we're investigating." Something with some scientific validity.

Zwicky didn't believe in UFOs either. After spending a decade studying the skies with telescopes, he was familiar with the tricks the atmosphere can play with light as it streams down on Earth. He told O'Mara that UFOs were nothing more than atmospheric disturbances, which could be caused by a variety of conditions. Which is just what the Air Force told the public. It's not known who else O'Mara consulted to formulate his policy, but it's clear from their letters that he regarded Zwicky as both expert and ally against the "frustrated socialists" that he felt dominated America's professorial ranks.

Zwicky's diaries show that he also spent time in the early postwar period at Fort Detrick in Maryland, working on another secret military project. At the time, Detrick was the center of biological warfare research in America. Some Operation Paperclip scientists from Germany were also involved in the research there. Zwicky's notes show he worked on "an experimental airborne infection" program, using rabbit fever and melioidosis. Though the diary entry is tantalizingly brief, it's now known that at the time, the United States was researching just such agents for use in wartime. Melioidosis is a bacterial infection, producing fever and pneumonia-like symptoms. It is endemic to places like Vietnam; because of its long incubation period, American soldiers who contracted the disease in Southeast Asia referred to it as the "Vietnam time-bomb."

Rabbit fever, or tularemia, also causes fever, fatigue, and headaches. Like melioidosis, tularemia is not necessarily fatal, but if what you want is to incapacitate your enemy, both are effective. As the chill of the Cold War spread across Eastern Europe, the Soviets were conducting similar research. Some historians claim Stalin's biowarfare program was behind an outbreak of tularemia among German soldiers at the battle of Stalingrad in 1942.

One of the professor's ideas that the military was most interested in was one that would transfer his inventions in jet propulsion to the underwater medium. "Problem: how to make a jet engine work under water," wrote *Time* magazine.[3] "Solution: make it use a fuel that burns in water instead of air." According to the magazine, Fritz Zwicky, "astronomer, physicist and rocket expert of Caltech has developed such an engine for the U.S. Navy, which presumably wants to use it in torpedoes (or) in anti-submarine devices."

The motor was an outgrowth of Zwicky's work at Aerojet during the war, and the Navy had become so enthusiastic about it by this time that it refused to allow him to answer a single question. "It has also warned Aerojet Engineering Corp. of Azusa, Calif., which is working on the device, to keep it quiet."

With a little sleuthing at the US Patent Office, *Time* uncovered enough information about patent no. 2,461,797 to give a pretty fair description. The underwater jet would work much like a ramjet, utilizing a single cylindrical chamber with openings at both ends. As water came in the front of the tube, it was mixed with a water-reactive propellant. A special device called a surface tension depressor helped mix the fuel. The list of potential fuels included metallic potassium, sodium, white phosphorous, and various metallic hydrides.

When such fuels hit water "they decompose it violently by uniting with its oxygen, giving off heat and a large volume of hydrogen gas." The explosive mixture was shot out the back "as a high-speed jet. The reaction from this drives the engine (and the torpedo) forward."[4]

Despite his work on secret defense projects, the Swiss professor's chief extracurricular interest at the time was a more lofty one: Like Frank J. Malina before him, and Robert Goddard before Malina, Zwicky dreamed of piercing the fragile envelope that since the beginning of time had sealed humankind in a safe but confining compartment. As it happened, he had just the tools to make the attempt: the German V-2 rockets.

After the war's end, Operation Paperclip brought 600 German scientists from Peenemünde, including 118 rocket experts,[5] to Fort Bliss in Texas, just fifty miles down a desert highway from the White Sands National Monument in New Mexico. The product of a shallow sea that covered southeastern New Mexico 250 million years ago, the remote, inhospitable, and startlingly bleached gypsum sand dunes had already seen a good bit of history; Sheriff Pat Garrett hunted down and killed Billy the Kid there, and one of the deadliest shootouts of the Old West took place in the little town of Lincoln. Now, the Southwestern desert was about to become home to America's budding postwar rocket program.

In 1945, the US Government created the White Sands Proving Ground, which later became the White Sands Missile Range. The facility covered 3,200 square miles, making it almost as large as Rhode Island and Delaware combined. The military shipped the hundred or so V-2s it had confiscated in Germany to the Proving Ground, where the expat German engineers taught their American counterparts how to operate the bulky, 46-foot-tall, liquid-fueled rockets.

The V-2 was too bulky and underpowered to reach space, but Zwicky thought that if he put some explosives in the nose cone, along with some bits of metal that he called "manmade meteors," some of those slugs might just make it. If the charge was ignited at the peak of the V-2's trajectory, the atmosphere would be so thin that some of the slugs might "assume the appearance of meteors" and escape the Earth's gravity, he wrote.[6]

On the basis of his military contract work, Zwicky was given the go-ahead by the Army Air Force's Ordnance office. The project caught the fancy of journalists around the country. Until then, nobody and no thing from Earth was known to have escaped the planet's gravitational pull. The reporters wrote that the launch would be a great first step on the road to space travel.

"In searching for the key to space travel, we first throw a little something into the skies, then a little more, then a shipload of instruments—and then ourselves," Zwicky told the journalists.[7] Later on, he would add a fifth stage: "Follow up the invasion of interplanetary space by an attempt to reconstruct the solar system so as to make the planets and their satellites habitable by man."[8] But he wasn't yet ready to spring his more fanciful ideas on the world.

Despite approving the professor's project, the military and the government still had little interest in space. Knowing this, Zwicky couched his proposal in strictly pragmatic terms, saying the experiment would reveal important information about the upper reaches of the atmosphere. Such information was critical to addressing "problems of supersonic and hypersonic aerodynamics through a large range of air densities." That was something the military could sink its teeth into.

Zwicky was correct that, at the time, the upper atmosphere was still something of a mystery. Some scientists even believed radiation belts surrounding Earth would make human space travel impossible. Space was so exotic that some people, even the educated, worried that puncturing it might blow open some sort of Pandora's box, out of which would pour countless evils. Zwicky tried to reassure the worriers. No one on Earth would be put in danger, he said. If any of his meteorites fell out of orbit, they would simply burn up on reentry.

His enthusiasm was not shared by colleagues at White Sands. Their number included two leading scientists, James Van Allen of Johns Hopkins University and Fred Whipple of Harvard. Whipple was outspoken about Zwicky's proposal, calling the mission a stunt. He wanted to charge the Swiss professor for the waste of taxpayer dollars.

The public, however, was enthusiastic about the launch of the Zwicky Meteorites, as some started calling them. If the experiment works, "it will open a thousand doors to invaluable scientific research," wrote *Science Illustrated*.[9] Zwicky said his next step would be to "pelt the moon." Citing Zwicky's accomplishments in supernova research and his war work, the magazine said that even if "initial attempts fail, it is anticipated that further trials will succeed; Professor Zwicky has the habit of success."

The space shot was set for December 17, 1946, using a V-2 that had been given to Johns Hopkins University. This was likely the source of Van Allen's displeasure. At the time, each V-2 was jealously cherished by its possessor. In any case, the rocket lifted off just after 10 P.M., reaching an altitude of 114 miles. At first, it seemed a success. Yet none of the thirty cameras and telescopes trained on the rocket detected any of the planned ejections. The explosive charges failed to ignite. The Swiss professor

Zwicky's first attempt to penetrate space occurred on December 17, 1946, when he launched a German V-2 rocket from the White Sands Proving Ground in New Mexico. The failure of the launch prevented him from getting another chance for eleven years, until Sputnik appeared in the skies, alarming the West and giving birth to the space race.

blamed last-minute fiddling by the launch team but assured the public that he would try again soon, and this time would succeed.

It was not to be. An angry Whipple said no further space shots would be done without major changes in the experiment. Zwicky sniped back, calling Whipple's response a "sad one." However, Zwicky appreciated the man's willingness to state his case openly, "most commendable when compared with the current habit of most American reviewers of scientific articles [to] shoot from ambush," using "methods which are usually only thought of as being practiced in police states."[10]

It took eleven years, and a national crisis brought on by the launch of the Soviet Union's Sputnik, before the professor got his second chance.

In the meantime, Zwicky was called to military duty once more, this time by the Swiss. With the chill of the Cold War settling over Europe, with countries on its borders falling under the Communist heel, Switzerland became increasingly anxious about Soviet intentions. Aware of what Zwicky had done for the Americans, the Swiss military invited him in early 1947 to study their defenses and recommend appropriate measures to counter future Soviet adventurism. This was a job for which the imaginative Swiss professor was particularly ill suited. When his imagination was untethered from practical considerations, his ideas could range deeply into the realm of fantasy. Such was the case with his proposals to the Swiss. Among his suggestions, parts of the modest Swiss air force should be stationed on floating runways in lakes. Jet fighters would be launched from submarine tunnels in the rock.[11] The Swiss government was so baffled they didn't know what to say. Some thought the astrophysicist had gone off his rocker. His homeland politely thanked him and hoped he would go away.

Zwicky was humiliated by the affair, and this undoubtedly played a part in his later denunciations of his country and its inhabitants as stupid and shortsighted. Zwicky already felt a secret shame about the canton of his youth, Glarus, because it was the last place in Europe to put a "witch" to death.[12] The 1782 decapitation of poor Anna Goldi was a long time ago, but he still considered it a sin that somehow belonged to him two centuries later.

Despite his failure in strategic planning, the trip had a consequence so profound that Zwicky's life would be completely changed, bringing a sense of lasting peace that he had not known since that childhood train ride along the Danube. Traveling in central Switzerland on one of his appointments, he took a room at the Hotel Falken in Thun, on the picturesque lake of the same name. A pretty blonde girl of 18, Margarita Zürcher, was behind the desk one afternoon, snickering with a colleague over the fantastic stories the American professor had been telling. About how people would soon be flying to the moon in rocket ships.

"I couldn't stop laughing," Margrit recalled many years later, in an interview. "But then I noticed he was sitting in the lobby. My colleague said that he may have heard us laughing, and perhaps I should go out and apologize." She did, "and then we got to talking, and that was the start of it."[13]

Zwicky, only months from turning 50 years old, promptly asked the girl on a date to go hiking. Showing the spunk that she would demonstrate over the course of her life in America, Margrit accepted. A most unlikely romance bloomed. Margrit may not have been put off by the age difference, but her parents were less pleased. They had greater plans for their daughter than to go off with a strange man from the United States, even if he were Swiss. To most Swiss at the time, "America was as far off as the moon," said Franziska Pfenninger,[14] Zwicky's middle daughter. To convince Miss Zürcher's family that he possessed the resources to provide for Margrit, he purchased a house on Lake Geneva for her and her family.

Three months after they met, the couple married. When Zwicky returned to Pasadena with his young bride, local society was not so much scandalized as dumbstruck. "The American ladies were shocked that he picked a young chicken and brought her back without telling them," Fran, as she is known to friends, said. It was common knowledge around Caltech that Fritz's first marriage had failed, a fact that made the robust Swiss professor a catch in his educated circles. Margrit was formally introduced at a campus social event. "All the ladies kept saying to mom, 'let's go powder our noses,' Fran said. "She didn't go because she knew they would ask her all these nasty questions."

Fran is a charming, friendly woman with cropped hair and a frank, unflinching manner that is an echo of her father's blunt candor. Born in 1950, she lives in a small, comfortable apartment an hour's train ride from Zurich, surrounded by keepsakes from her past. There are primitive Canadian tribal paintings that she bought with her doctor husband,

Zwicky, in his characteristic bolo tie, with his second wife, Margrit, in about 1965. Though he was three decades older, they made a long and happy marriage, producing three daughters. Around this time, Margrit began helping him with his work, searching old photographic plates for supernovae.

Hanspeter, who died in 2012. After his passing, she gave up the big house on the hill and moved nearer the center of Mettmenstetten, a 900-year-old bedroom community south of Zurich. Along with the paintings and assorted family pictures of Hanspeter and her two grown children are treasured mementos of her father's, his stamp collection, and the heavy dining room table on which she served coffee one cold winter day in 2015.

The table is an antique, originally owned by Fritz's grandfather, who raised him. It's made of walnut and slate from the mountains surrounding Glarus, the first peaks her father climbed. Stone from those mountains was also used to construct the cantilevered façade of the Cahill Astronomy Building at Caltech. The same mountains provided the stone for Zwicky's grave, which lies alongside the heroes of the 1338 battle of Näfels.   .

Within months of their wedding in October 1947, the one-time bachelor had settled into the placid routines of married life as if he had been waiting for it his whole life, and perhaps he had.

"Margrit has transformed my bachelor flat and made it spic and span," he wrote Rösli in December. She did the same for his cabin at Palomar, which was not easy, because it snowed six inches on their first visit. He was starting life fresh, he said. At last, he felt ready to return to his personal and international work with a renewed energy. Margrit, too, was very happy, "if I am not mistaken."[15]

Eleven months after their marriage, and less than two weeks before the birth of his first daughter, also named Margrit, who was known as Margritli in the family, he bought a simple bungalow at 2065 Oakdale Street. It was in a modest neighborhood within walking distance of Caltech. Margrit started English classes at Pasadena City College; within a few years, her command of the language was equal to her husband's. Her accent eventually disappeared, unlike his, which hung on stubbornly.

She fitted herself willingly into the conventions of an academic life. Never retiring but at the same time never aspiring to be anything beyond wife and mother, she had but one shortcoming, in the beginning. Her familiarity with food consisted almost exclusively in the consuming of it. Her girlhood plan was to work in the hospitality industry, not the kitchen. "She became a good cook, but in the beginning father would take her out three or four times a week," Fran said. "Mom thought he enjoyed taking her out and showing her off. The truth was, father wanted a decent meal."

When they went out to eat, it was often to the one Swiss restaurant in Los Angeles, operated by the Moser family, near the Los Angeles Memorial Coliseum. Like much about Fritz Zwicky, excepting his imagination and tempers, his taste in food was simple and unadorned, running to pot roast, beef stew, and onion and cheese pie. "We didn't live on a big foot" is a Swiss expression his oldest daughter, Margritli, uses to describe his approach to family life.

While the former Miss Zürcher stayed mostly in the background at society gatherings and scientific meetings, she was no shrinking violet.

Her sharp, "Fritz!" was the one command that could cause him to stop and withdraw from a public spat, whether it was with a colleague or the maître 'd at a restaurant.

She also knew how to put the unexpected, and uninvited, suitor in his place. One story her daughter laughingly told centered on the young professor who flirted with her during an elevator ride in Caltech's Robinson building, where her husband had his office on the second floor. She was wearing stacked heels and a short skirt and doubtless made a delicious sight, with her blonde curls and wide smile. When she said she was married to Fritz Zwicky, the man recoiled as though he'd been slapped. Whether it was because of his deep respect for the institution of marriage, or because he had no desire to tangle with Zwicky, is unknown.

An early poem Zwicky wrote for his young wife shows how deeply he bonded with the girl from the lake country. "Grow strong, my comrade / That you may stand unshaken / When I fall, that I may know / The shattered fragments of my song will come / At last to finer melody in you."

Fritz Zwicky was not an elegant writer, but it's hard to argue that he was not a determined romantic. The martial theme he struck was a common one for a man who believed that he was in a constant battle against forces and people he considered enemies of human fulfillment. Or, at least, his.

Once back in Pasadena, still nursing a grudge against people like Feynman, who seemed to have lost faith in the ability of science to solve big problems, Zwicky took on new challenges to prove that a scientist need not shut himself away in the laboratory. Knowing reconstruction must follow war, he determined to restock the scientific libraries of educational institutions in the war-torn nations. Over the next decade, the Committee for Aid to War-Stricken Scientific Libraries, which consisted mainly of Zwicky himself, boxed up more than 100 tons of books for shipment to South Korea, China, the Philippines, Taiwan, France, and Germany. Leaning on his military contacts, he obtained the help of the Navy in sending the books abroad. The service was only too glad to help, since Zwicky was at the time researching undersea propulsion systems for them.

The ever-present local journalists in Los Angeles were happy to publicize the effort, taking pictures of Zwicky surrounded by piles of books. "We feel pretty good today," he grinned, speaking to reporters, in

June 1953. He had "already shipped 1,500 boxes" and had just received a grant from the Ford Foundation to ship tons more to Chiang Kai-shek's forces holding out on the island of Formosa (now Taiwan). A grateful Heinrich Lübke, president of the Federal Republic of Germany, said the shipments to his nation, humiliated by defeat in war, kindled "the first new hope which aided us to overcome our spiritual isolation."[16]

Less pleased were Zwicky's bosses at Caltech, who complained that he had used astronomy department funds to obtain the books, and that he did it all without asking permission. "This was quite gracious, but it cost us a lot of money, which we didn't have at the time," his department head, Jesse L. Greenstein, said. The objections were kept from the public. Once the effort had begun, they could hardly stop it without looking small-minded.

Zwicky also plunged into a campaign to relieve the suffering of European families victimized by the war. As a volunteer for the Pestalozzi Foundation,[17] which was formed in 1944 and built housing in Italy, France, and Germany for children orphaned by the war,[18] he raised hundreds of thousands of dollars. He was so successful at raising money—$258,943.29 in 1950 alone—that he was chosen chair of the organization's Board of Trustees.

This pattern was repeated. With the same confidence and lack of self-consciousness with which he approached exploding stars and jet propulsion, he leaned on his colleagues to step up and give money to worthy projects. After becoming vice president of the International Association of Aeronautics some years later, he was even able to squeeze donations from the Soviets, at a time when they refused on general principles to contribute to other decadent Western organizations.

It's true, however, that his participation came with special challenges. In a letter, the founder of the Pestalozzi Foundation, H. C. Honegger, chastised Zwicky for driving off donors. Honegger apologized if his report would spoil Zwicky's appetite, but "since you are over-eating as it is, losing a little weight may do you good."[19]

It was true, Zwicky's eating habits and lack of exercise had brought on the weight gain that often afflicts people in middle age, when too much work inspires too much indulgence. He had plenty of opportunity to indulge, since Zwicky's reputation for wit and clarity as a public speaker was so well-known that he constantly received invitations to give keynote speeches around southern California to both scientific and civic groups. In early 1947, he told the local Rotary Club in Azusa that the key to discovery was to "work on the principle that the aspects of life are inexhaustible. Any number of wonderful things remain to be discovered. Therefore, in basic things, believe nobody." This was a distillation of his by then settled attitude that his accomplishments in space and war were a product, not of any special genius that belonged to him alone, but of his fierce independence.

The secret to survival in science, he said in another talk, was "not to be scientific about colleagues, or, especially, superiors." The fact that these attitudes alienated potential allies didn't alter his course. For Zwicky, being the truth-teller who fearlessly fought convention was so deeply ingrained that he could not consider being anything other than the rough micro-Zwicky of von Kármán's imagination.[20] That an entire generation was being born—the baby boomers—who would adopt many of those same principles, rebelling against authority, coining the term "generation gap," and following their own dictates, would come as a surprise to him. An unwelcome one.

This instinct for seeking out his own, personal, truth had by then evolved into a philosophy that he was ready to reveal to the world. An occasion presented itself when he was invited to give the Halley Lecture at Oxford University in the spring of 1948. Inaugurated in 1910 to coincide with the return of Halley's comet, the Halley Lecture is one of the laurels crowning a career in astronomy or physics. When Zwicky boarded the Queen Mary, which had been restored to its art deco grandeur after serving as a troop ship during the war, he likely harbored doubts about the wisdom of touting Morphology before the cream of the scientific world.

But the moment seemed ripe, even if the Halley Lecture was usually devoted to such issues as "The Origin and Structure of the Solar System Comet Cloud," the topic of the 2016 lecture. Zwicky believed in his process, and he believed it was at its heart deeply scientific, since it was based on the idea of considering problems from all points of view, without prejudice. And he wouldn't have been Fritz Zwicky if he had shrunk from taking the leap at Oxford on May 12.[21]

Morphology was concerned with "the structural interrelations among objects, among phenomena, and among concepts. Because of this stress on completeness," he wrote later, "the indispensable requirement for morphological analyses and constructions is the complete absence of prejudice."

To solve a problem, Zwicky believed that one must first understand its shape. In other words, the limits of the investigation must be established, even before investigating.

This was not a new idea. Johann Wolfgang von Goethe had proposed the same theory of knowledge 200 years earlier. Konrad Rudnicki, a Polish astronomer and priest who was for a time a collaborator of Zwicky's at Caltech, was one of the first to realize that Fritz Zwicky was essentially a Goetheanist. When they met, Rudnicki cautiously suggested as much. "Of course I am [a Goetheanist], but I do not talk much about it," Zwicky replied. "They regard me [as] crazy anyway."[22]

Today, Goethe is considered a poet, as well as the originator of self-help aphorisms such as, "Knowing is not enough; we must apply. Willing is not enough; we must do." But, according to Rudnicki, Goethe considered himself a scholar writing verse in his leisure.

Most theories of knowledge are based on Kant's sharp distinction between thinking and perception. Through interaction between the two, an understanding of an object or event, the "thing-in-itself," comes about. Different theories of knowledge lead to different conclusions, from the idea that the "thing-in-itself" is unknowable to the idea that we can obtain more and more knowledge about it in some way. "According to Goetheanists, a theory of knowledge must not rely on any research discipline, including logic. In essence, this theory of knowledge is pre-scientific," Rudnicki wrote.

Under Goethe's theory of knowledge, thoughts are treated as perceptions. Readings from measuring devices are perceptions, too. The whole of reality, then, is solely constituted of perceptions. Since everything consists of perceptions, there is no reason not to trust any particular set of thoughts, or to raise "objections that they distort reality." In essence, this strips away all preconceptions like layers of paint being removed from old furniture, revealing the essence of the object beneath.

To Goethe, and to Zwicky, misinterpreting phenomena—like deciding the sun goes around Earth because that is how it appears to us—is a much less dangerous problem than deciding arbitrarily at the start that some phenomena are not real. In other words, preconceptions are a greater danger than misperceptions.

The conclusion that the whole of reality is made up of perceptions has important implications. A common way of doing research is to make a hypothesis. If the scientist discovers something that contradicts his or her hypothesis, she has to account for it, either by introducing a new parameter or admitting the original hypothesis failed. "Today, we use paradigms . . . based on the idea that if a given procedure yielded good results in the past it will do so again," Rudnicki wrote. "There are thousands of research papers appearing every year that we know will within a few years be proven false, yet degrees, fellowships and scientific titles are awarded on their basis."

Zwicky might have written that; he believed the scientific journals were filled with garbage. The great problem, which Zwicky saw and which Goethe described, was the process of making a hypothesis is not subject to any rules. Zwicky's favorite way of describing this was to envision rain falling on the ground. As the water runs to the sea, it makes ruts in the earth. Over time, those ruts grow deeper and deeper. Meanwhile, the area outside the rut goes unexplored. Similar to the rain opening up the earth, "ideas unlock the doors to aspects of life, fixing the attention of men on some aspects, ignoring others," Zwicky said.[23] Once in a rut, the investigator tends to simply dig deeper. "He does not even take the excavated debris with him like the waters, but throws it over the edge, thus covering up the unexplored territory and making it impossible for him to see outside his rut.

"It is these two major requirements, visualization of the true facts of the world, and detachment from all prejudice, that are so exceedingly difficult to achieve," Zwicky argued. "While the common man is generally quite modest in his claims as to the extent of his knowledge, and his unbias [sic], scientists just as generally harbor the illusion of knowing it all and thinking and acting objectively. . . . In particular the United States of America has seen many of these characters performing high trapeze acts of gigantic deception."[24]

Morphology "also permits man to forge his way into any special field of knowledge . . . and confidently confront any so-called expert or specialist in this field."

Zwicky was a tireless advocate of his invented theory of knowledge, which he called "Morphology." Here he is shown lecturing on the subject at a conference in Athens, in 1970. He claimed the technique, which considered problems from every possible vantage point, without any preconditions, enabled him to make his fundamental discoveries in jet propulsion, exploding stars, dark matter, and rocket fuels.

He was thinking of his work in astrophysics and jet propulsion. Employing morphology, "rank outsiders . . . have produced results during the past decades" on the chemistry of propellants, on astronomy and in the science of war, "which no specialists in those fields even dreamed of."

Goethe first advanced this theory of knowledge, but he never attempted to systematize it so that it could be used by anyone. Zwicky attempted to do that with what he called his Morphological Box, which was like a file cabinet. As *Fortune* magazine described the idea in an article that treated the Swiss professor like a prophet of the new scientific age,[25] the box treated every problem in three dimensions. For instance, a morphological box constructed to investigate new combustion engines would consider potential propellants, the motion of engine parts, and means to increase thrust. "Pull out any drawer in the file, and in each compartment reposes a particular engine or one still to be invented," *Fortune* wrote. The article suggested that morphological thinking could be key to solving humanity's greatest challenges, "from new power plants employing atomic energy to radio-electronic devices to cancer research."

Zwicky's advocacy of morphology bore some similarities to other utopian philosophies arising in that era, from Ayn Rand's Objectivism to L. Ron Hubbard's Scientology. And, like the others, Zwicky was a determined proselytizer. Morphologists in the future—"who understand one another and who do not necessarily need to be organized"—were sure to "play a decisive role in the realization of the genius of man." Various attempts were made in the ensuing years to organize chapters of morphologists. A first symposium on his system was arranged at Caltech in May 1967 by the International Society for Morphological Research. Even earlier, in 1951, the Air Force was interested enough in the approach that it asked him to submit a proposal to teach it to their officer corps. The Swedish Morphological Society is still in existence.

But it never caught on the way other movements did, partly because few people were able to understand the subtler aspects of the process. Rudnicki admitted there was an additional problem: Morphology was particularly adept at giving a general idea where a solution lies, but it often could not produce an unequivocal answer to the problem at hand.

Von Kármán cited an example of the philosophy's limitations. During the Korean War, Aerojet received a large order for smokeless jatos. When yet another problem arose with the solid propellant, von Kármán suggested bringing Zwicky in. The Swiss morphologist said he would need at least six months to "establish the morphology" of the cracks that were making the propellant unstable.

A few days later a junior engineer got the idea of putting slots into the propellant so that it could expand and shrink with the temperature without cracking. It worked. "Sometimes it is better to seek a simple engineering trick to solve a problem than to indulge in elaborate analysis," von Kármán said.[26]

Even so, Rudnicki considered Zwicky's morphological method preferable to the way science was done then and is still done now.

Zwicky's Halley lecture received an enthusiastic reception, which encouraged him to spring it on the public at large when he returned to America.

"A master key to the solution of the world's greatest problems—a scientific thought technique designed to keep this nation always five years ahead of any dictatorship like Russia—has been evolved by Dr. Fritz Zwicky, it was learned yesterday,"[27] wrote the *Los Angeles Times*. The writer fretted over trying to describe the method in a newspaper, comparing it to Einstein's unified field theory, which, of course, did not exist. Still, he gave it a try, saying the technique "aims to solve problems only after all possible solutions have been worked out in advance and systematically sifted." Through its use, the writer said, Zwicky had already produced ideas for 576 types of new jet engines. It was an absurd number. But the *Times* science writer, William S. Barton, did not challenge it. Over the years, he went to great lengths to stay on Zwicky's good side, even helping him fill out his stamp collection by scooping up new issues on his travels.

"Morphology is a technique only free men can use," Zwicky said. "In fact, the morphological method and our ability to use it is the most important advantage we shall have in any war with Russia." In his mind, the method was the solution to every problem America faced, from Stalin, to atomic weapons, to runaway population growth. "These three problems, alone, are too vast for the average man to cope with as he staggers about under high taxes and seeks to maintain a happy home," he said.

The once-young rebel from the mountains of Switzerland was now 52 years old and chafing, like many heads of household, at the evils of big government. The fact that his income was many times that of the average working person did not, in his mind, justify the sour stomach he got when he came home at night "to find income tax blanks waiting."

In his article, Barton fulfilled his journalistic duty by running Zwicky's theory of knowledge past an unnamed "noted astronomer," who damned the idea with faint praise. "Fritz has one of the most penetrating minds of his time, but this makes him too impatient when we disagree with him," the astronomer said.

As for Fritz's current research at Caltech? "Right now, he is speculating upon the possibility of harnessing the moon, and driving it, with the aid of atomic energy, through the heavens," Barton wrote.

Yes, he did say that. The postwar exuberance that had infected the nation gave Fritz Zwicky a full-blown case of the mental bends. After taking the world out of the coal and iron-horse age and into the age of jet travel, science could do anything, he and many others still believed.

It was common for Fritz's friends and admirers to say that he could not possibly have meant half of what he said. For instance, turning the sun into a spaceship, dragging its planets behind like ducklings, to explore other galaxies. Or solving population growth by lassoing another planet or asteroid and towing it closer to the sun to make it habitable. The apologists were wrong. He did mean those things. They were as much a natural product of his theory of knowledge as his inventions in rockets.

"In actuality," he wrote in his book, *Morphology of Propulsive Power,* "the reconstruction of the solar system is inevitable and necessary, not only because man's curiosity and initiative do not stop at anything, but also because of the urgency of finding large-scale solutions for the problems arising from the overpopulation of the Earth and because of the possible advisability of separating potentially or practically incompatible groups of men and ideologies."

By this time, Zwicky had developed a more bitter view of human nature. Survival might require separating peace-loving morphologists from others, "Hell-bent on destroying themselves in some large scale nuclear catastrophe."

His admirers wanted to protect him from those who laughed at his wilder ideas. He pretended not to care. But of course it did bother him. It's just that, as time went on, he began to cherish his reputation as *enfant terrible,* even if he was long past the age. And the journalists, both in California and around the country, were only too happy to give him a megaphone for his ideas.

So he went right on making outlandish statements and proposing wild inventions. One that got more attention than any other was a device he called a terrajet. It was something straight out of the comics, a "mole-like missile" that could bore through the earth, taking rock and soil in at the front, reducing it chemically, and then ejecting it out the rear. That would push the vehicle forward, deeper and deeper into the bowels of Earth. "This is something you haven't seen yet, but which is absolutely in the scheme of things," Zwicky said.[28]

The terrajet could be useful in mining as well as warfare, he said. Countless mothers, who asked children digging in the back yard if they were trying to reach China, must have shaken their heads when they heard about it. At a press conference at the Beverly Hills Hotel, Zwicky said his machine could do just that.

The experts might have laughed, but it was Zwicky who got the last chuckle. The military soon contacted him, asking for the plans of his terror weapon.[29]

12

BRIDGES IN SPACE

ON A LATE SPRING day, less than three years after the world witnessed the terrifying power of the atomic bomb, a crowd of several hundred gathered in a mountain-top meadow to celebrate a technical marvel of science and engineering that was every bit its equal in imagination and daring.

With the 135-foot-tall dome of the Hale Telescope gleaming in the afternoon sun, James R. Page, chair of the Caltech Board of Trustees, addressed the crowd of dignitaries and professors. If this dedication ceremony were being held in England, he said, it would be attended by the pageantry of the state, "with its knights, heralds, pursuivants, kings at arms, admirals, captains, and of the church, with its bishops, priests and deacons, crucifers and choirs."[1]

To Page, this lack of pomp and splendor was not something to be ashamed of, but a matter of pride that fit the achievement. For the great 200-inch telescope inside the giant dome, grand as it was, was no bauble of empire but a workmanlike machine, built by a voyaging society still driving forward, into wilderness. That the wilderness it was penetrating was space did not make it any less remarkable than the ships of Isabella or the legions of Rome. Yet Page could not ignore the lessons of recent history on that June day in 1948.

"Today, in dedicating this telescope, we are face to face with the problem of the unpredictable consequences of knowledge," he said. "We cannot even guess what will come from this mighty instrument, or to what ends the fresh insights which we gain here will be employed.

"When the giant cyclotron was built at the University of California, nobody was thinking of the atomic bomb. . . . And yet that cyclotron contributed materially to development of one of the phases in the construction of the atomic bomb, just as this telescope may conceivably give us knowledge which, if we so choose, we can employ in the insanity of a final war.

"In the face of this dilemma, what is our proper course of action? Do we stop building telescopes? Do we close down our cyclotrons? Do we retreat to some safe, underground existence where we can barricade ourselves against our fears?"

No, he declared. "The search for truth is, as it has always been, the noblest expression of the human spirit. . . . The towering enemy of man is not his science but his moral inadequacy."

After that somber introduction, Ira S. Bowen, the director of the new observatory, outlined his hopes for the instrument. It was not built for a single purpose, he said, but to answer the great questions that lay beyond the reach of every other earthly instrument trained on the heavens.

A "great new power will be brought to bear on the study of nearly every astronomical problem," Bowen said, "whether it be pushing back the extreme limit of the observable universe or investigating in detail one of the earth's nearest neighbors."[2]

Bowen, the director at Mount Wilson as well, said the new telescope, with its 14.5-ton, 17-foot-diameter mirror, would build on the impressive work of the 100-inch Hooker telescope, which had dominated world astronomy for the previous three decades. Listing the achievements by Hubble and Baade at Mount Wilson, Bowen noted, "as often occurs, however, these investigations raised more questions than they answered. For example, is the universe really expanding or are the observed effects caused by some curvature of space?"

If so, then the universe, like a rubber band stretched to its limit, could eventually begin to contract on itself. This would make the universe a closed system. Instead of everything streaming away to infinity, at some distant time, things would begin contracting, when everything there is, and ever was, collapses on itself. Bowen's remarks showed that even then, the scientific world remained skeptical of the expanding universe.

"Up to now," Hubble told a writer from *Life* magazine, "we have only been able to brush space with our fingertips. Now we will be able to grasp it with our hands."[3] It was a nice analogy, showing that the great investigator could speak the layperson's language when he chose. Hubble wanted to give a talk at the dedication, but it had been canceled. It was the latest setback for the prickly astronomer, who had been stunned by Bowen's selection as the director of the new observatory rather than himself.

The decision was not about Hubble's undeniable achievements, but his character. His tendency to hog the spotlight alienated colleagues. "He has never put his mind on [the] somewhat subtle aspects of human relationships," wrote a critic.[4]

After his speech, Bowen put the dome, and the 60-foot-tall telescope nestled inside, through their paces. He rotated the reflector about its two axes, north-south and east-west, which would allow it to cover three-quarters of the night sky. The entire structure rumbled smoothly on its foundation, sunk twenty-two feet into the bedrock of the 5,600-foot-tall peak. It was so finely balanced, on thirty-two carriages, each with four wheels, that the apparatus was moved by a single, half-horsepower motor. A complete rotation took four minutes. For rapid motion or slewing, a two-horsepower motor was employed. Atop the thousand-ton dome rested two 125-ton shutters, covering the opening. Each night, they would slide open with a majestic rumble, revealing the patchwork tapestry of the heavens.

When the demonstration ended, the group disbanded, heading for their cars and the long trip back to Pasadena. It had been a simple, almost inauspicious ceremony, particularly when compared to the monumental effort it took to get there. Making the great Hale Telescope might not have compared to the generations of people whose lives were consumed building the pyramids, but it had been a momentous, error-plagued process.

Construction on the mountain began in 1935. The dome's outer structure, designed by the illustrator Russell W. Porter—who also made a memorable sketch of Zwicky in 1934—with help from von Kármán, was complete by 1939. Components were fabricated at the Westinghouse south Philadelphia plant and shipped by boat through the Panama Canal to San Diego, where they were trucked to Palomar, up rural road S-6, known locally as the Highway to the Stars.

The 200-inch telescope at the Palomar Observatory signaled the arrival of a "great new power" in astronomy. It was George Ellery Hale's grandest achievement, yet he did not live to see its completion.

Constructing the dome was a breeze compared to the task of making the mirror. Originally, Hale hired General Electric to fabricate the mirror from quartz. After $600,000 was wasted on the idea, it was abandoned. Then Hale approached the Corning Glass Works in New York, which was marketing a borosilicate glass called Pyrex. Pyrex required long heating times at very high temperatures to produce the very hard, low-expansion final product. It took a month alone to melt the glass for the mirror. The first try was a failure. A second attempt was made on December 2, 1934, as Fritz Zwicky's ideas about exploding stars began to capture the fancy of the public. The giant disk remained in the oven ten months, cooling only a few degrees each day, to prevent voids and cracking, before it was ready to be shipped west.

Finally, on March 26, 1936, the mirror began its sixteen-day trip by rail from Corning to California. Hale, always the showman, relished the idea of people gazing in wonder as his great mirror passed by. And they did, by the thousands, lining the tracks to watch the great chunk of glass roll by that would unravel the mysteries of the universe. Once it arrived in the Caltech optical shop, the next task began: grinding the mirror into its final concave, parabolized shape, a precise four inches deeper in the center than at the edge.

"It will be a comparatively simple task to grind the mirror," optics chief John A. Anderson told a reporter from the *Associated Press* at the time. "We hope to have the reflector ready by 1940." Hale was already dreaming bigger. He said it might be possible to make a 300-inch mirror, though he wasn't sure there would be that much more to see with a mirror that large. Today, reflecting telescopes in Hawaii have reached nearly 400 inches, while the Gran Telescopio Canarias in the Canary Islands has surpassed even that.[5]

Anderson was too optimistic. For the next eleven years, while a great war was waged and won, while the Depression waxed and ended, while George Hale sickened and died, and while Fritz Zwicky searched for supernovae and tried to pierce space, workers in the Caltech astrophysics laboratory scratched at the glass disk. Rightfully called the "moon shot" of the 1930s and 1940s, completing the mirror consumed even more time than it took to put a man on the moon after President Kennedy announced it as a goal in 1961.

Year after year, through the heat of summer and the misty chill of the southern California winter (work was halted for three years due to the war), Anderson's team ground away at the disk in a laboratory whose walls were lined with four inches of cork to prevent the glass from expanding and contracting with temperature changes outside. More than five tons of Pyrex were removed from the twenty-ton blank, including a final, delicate .005 of an inch near the outer edge.

At last, on November 18, 1947, the finished mirror, strapped upright on a flatbed truck and escorted by police with an entourage of newsmen, made a two-day journey south. At speeds between five and fifteen miles per hour, the mirror rumbled through small villages, past orange and lemon groves, and across canyons—the combined 40-ton weight of the glass and semi-trailer truck deformed one coastal bridge by nearly a half inch—before finally arriving at the mountain. The big truck carrying it had to be pushed up the last heights of Palomar by two other trucks. The layer of reflective aluminum was added on the floor of the dome, and then the mighty one-eyed beast was yoked to its 530-ton steel mount.

For Tom Gehrels, a young Dutch post-doc who came to America to make his career in astronomy, the sight of the Hale Telescope was the experience of a lifetime. It was "the greatest structure I had ever seen, more impressive even than the skyscrapers of New York," he effused.[6]

Zwicky himself did not attend the dedication that spring day in 1948. He was in Europe following his speech to the Royal Astronomical Society. Margrit, six months pregnant, was home in Pasadena. In two weeks, Fritz would board the Queen Mary for the return trip to New York. Back in Los Angeles, he would be robbed by a pickpocket on the Red Car trolley, an encounter that began to sour him on life in America.

In any event, he would not have been much interested in the ceremony, even if he were in town. He had been on the mountain for a dozen years, searching for supernovae and studying the structure of space at the helm of the 18-inch Schmidt telescope. The sight of the big dome rising from the cedars and pines was familiar to him. The gawkers who came to watch the construction work were a distracting irritation. Despite Zwicky's long familiarity with conditions on the mountain, Hubble was given the honor of taking the first "official" images, on January 26, 1949.

Two months later, a writer for *Time* magazine was invited to watch Hubble work, seated in the six-square-foot round steel cage, suspended seventy feet in the air. Focusing on the constellation Coma Berenices, he resolved everything ever found by the 100-inch Hooker in just six minutes.

"We reached one billion light years into space—twice as far as man had ever looked before. The tests confirm our previous conclusion, that the Hale telescope is an unqualified success," Hubble said.[7]

Real work at the Hale did not begin until November of that year, two months after the Soviet Union tested its first atomic bomb, sending shudders through the world's capitals.

By the time the Hale got to work, a dozen buildings had sprouted on the hilltop, including a powerhouse, cottages for the workers, and the Monastery, a half-mile away, where the astronomers lived while on duty. At first, there was no phone service. The complex was connected to Pasadena by shortwave radio. The dome of another new telescope had also risen. A 48-inch Schmidt camera, the big brother of Zwicky's favorite instrument, would scout targets for the big Hale Telescope. The new Schmidt was ideal for the work, because its 14-inch photographic plate covered thirty-six square degrees of sky. With a thousand plates, it could search the entire northern sky. The National Geographic Society was funding just such a survey, an Audubon bird count of the heavens, which would take four years. Doing the same with the deeply penetrating but narrow field of the Hale Telescope would take 5,000 years.

The writer David O. Woodbury visited early in 1949 and produced a report for *Collier's* magazine that conveyed the excitement of a nation flush with confidence after military victory and economic resurgence. "This is an exciting time to be alive," Woodbury wrote.[8] "Yesterday: the coming of atomic energy; today, the first message from continents of stars so frighteningly far away that they may actually be at the ends of space."

According to Woodbury, human beings had long taken comfort from the apparent permanence of the stars in the sky. Zwicky and Baade, along with others, had shown how impermanent they were. Palomar would go further, casting human beings up on the most distant shores of the universe, posing questions lying in the realm of theologians and philosophers.

The scientists working at Palomar were not concerned with such questions, however. They were there to look, to find, to measure, and to tell the world about it. Their explorations might not expose them to the physical danger of earlier pioneers. There would be no long months on storm-tossed seas, just hour after hour suspended in a small cage, high above the observatory floor.[9]

Woodbury described a typical night at Palomar. The astronomer and his assistant, "quite natural, everyday men," travel to the mountaintop by car. The last section of the trip required a guide dog's sensitivity, since streetlamps and headlights were forbidden on the mountain. After entering the vast, echoing hall of the observatory, the men discussed the job at hand. Then the astronomer ascended to the observation station high "in the throat of the gigantic machine."[10] The assistant headed to the organ-like console that steered the massive telescope to the correct position for the night's viewing.

"The skylight fades. Within the dome a few heavily shaded lamps spill pools of yellow on the floor. Up into the crisp, dim air the mighty silhouette of the giant reaches, utterly quiet, waiting." The astronomer speaks just a word or two into the intercom, and the telescope glides into position.

"The gentle thunder of the moving dome rumbles briefly and is still. Into the field of view slips a guide star; the astronomer fiddles with his controls to center the crosshairs of the eyepiece, slips a glass photographic plate in place"—no one uses glass plates nowadays—"and pulls out the slide. Light thousands of years old begins soaking into the emulsion. With the target focused, the great machine automatically follows the star as it crosses the sky" with Earth's rotation, at 800 miles an hour.

For over an hour, there was no sound but the "thin, faint hum of the driving motors that keep the telescope dead on its objective." Finally, the plate is removed and the giant is on to another target.

It went on like that all night, or until the work was done. At last, the observer, stiff and chilled despite his warm parachute suit—Margrit made Zwicky special gloves with only the ends of the fingers exposed to prevent frostbite—the astronomer returned by way of the elevator platform. Then he was off to the darkroom to develop the heavy glass plates and put them in a drying rack.

One of the most important tools astronomers would rely on at the Hale was the spectrograph. These instruments slice up light into its constituent parts, producing a rainbow-like spectrum that, properly interpreted, reveals the temperature, pressure, velocity, and chemical composition of the substances from which a star is made. A solid gives off a different rainbow than a gas. Elements show up as dark lines at specific places between the red and blue ends of the spectrum.

Astronomers at Mount Wilson had used a spectrograph to find out what the sun was made of. But a good spectrographic image requires much more light, and enormously more time, than a photograph. A low-dispersion, faint-object spectrograph breaks light into about a thousand colors, while a photographic plate with a filter looks at only a fifth of the wavelength range.[11] This limited the Hooker's ability to say much about stars in other galaxies. One of the spectrographs at the Hale was thirty feet long and weighed many tons. These instruments would allow scientists to discover whether the stars in our section of the universe were like those in other territories, a crucial piece of evidence if scientists were to ever draw conclusions about the makeup and personality of the universe as a whole.[12]

At last, with the sun up, and after developing his plates, the astronomer headed for a warm bed in the Monastery, the dormitory-like building in an oak grove. The Monastery lounge was where the astronomers gathered when they felt like socializing. There was a small library. At Mount Wilson, Hubble, ever anxious to prove his superiority, boned up on obscure encyclopedia articles so that he could expound on arcana to impress distinguished visitors over dinner. But at Palomar, Woodbury was surprised to find the scientists, like ordinary men and women, favored light reading—detective novels and popular magazines. Occasionally, a game of poker broke out when the mountain was shrouded in fog or snow. No one played chess. Back in the dome, the radio was a companion. Although most of the astronomers liked symphonic music, there was little to be had on the mountain. Popular music on the hit parade was their primary accompaniment as the Big Eye rumbled along on its nightly explorations.

One night, a newscaster announced that the Soviet Union and the United States were in fierce competition to see who would dominate, not just Earth, but the universe. "I hope," the astronomer on duty called down to his assistant, "he'll remember to let us know how it turned out."

Zwicky stayed sometimes at the Monastery, but when Margrit accompanied him, he lived in a rustic cabin he purchased in late 1947, when a half-foot of snow lay on the ground, and the mountain peak still belonged to him. The only heat was provided by a wood stove, which made getting up in the morning and making coffee an uncomfortable duty. It would be several years before he wired the place for electricity.

During the winter, most of the astronomers trekked to the observatory by snowshoe. Not Zwicky. He skied back and forth, showing off the skills he developed as a boy in the Swiss mountains.

Time on the new Hale instrument was so dear that even the most accomplished stargazer rarely got more than a few days each month at the controls of the giant telescope. It was some time before Zwicky got his first, official chance at the Hale, a fact he deeply resented—and considered proof of the prejudice against him. Which it likely was.

No exotic animals were slain, no strange plants bottled up and stored in the hold of a frigate. Heavy glass plates were the only proof of the journey into the darkest, most remote territories of the cosmos. But the information they carried would support, or upend, long-held theories and beliefs.

"Nobody can grasp the tremendous space now open to us for study without feeling a great deal less responsible for the behavior of the universe and a great deal more responsible for his own behavior," Bowen told John K. Lagemann, another *Colliers* writer.[13]

Lagemann recorded the impressions of a laborer who had worked on the big telescope. "Like we were all shipwrecked here together in an ocean of space," the man said, looking up at the mighty instrument. This man knew something of the subject. During the recent war, his transport ship was torpedoed off Guadalcanal. The impression at Palomar of Earth's place in the cosmos reminded him of the way he felt while clinging to a life raft in the Pacific. "It doesn't feel much bigger, either," he said.

Most of the astronomers were older, often grandfathers, earning between $5,000 and $8,000 a year. What was behind their excavations in the stars? "Sheer curiosity," said Hubble. "It's the basis of all science."[14]

Unlike most other sciences, where the adventurer might sort through piles of information from a variety of experiments, all astronomers had to work with were the faintest shards of ancient light, which they divided and divided again, teasing out the secrets of distant stars and galaxies. It wasn't easy. Data from a single exposure could take weeks to analyze.

Over the next few decades, the 200-inch telescope would revolutionize our understanding of the lives and violent deaths of stars, help untangle what made them tick, and cut a sure path through the underbrush to the ultimate discovery of the Big Bang when, and where, everything started.

Zwicky's former partner—now bitter enemy—Walter Baade would use it to firmly establish the idea that there are several distinct populations of stars, a diversity of species and races not unlike the genetic history of humans. Some stars, known as Population II, were old and simple-minded, made predominantly of the most basic elements, hydrogen and helium, that were around when the universe began. Others, the Population I group, were younger and far more dynamic and interesting. They were composed of heavier elements like carbon and calcium, the seeds of life on Earth and, perhaps, elsewhere. It was a puzzle where the younger stars got those elements, until scientists building on Baade's work realized they must have come from the older stars. The deaths of older stars essentially fed the younger star group with more complex elements, including those necessary for life on Earth.[15]

Baade made another discovery with the Hale that would ensure his lasting fame. While trying to measure the distance of variable stars, like Cepheids, he discovered that the Andromeda galaxy was twice as far away as previously thought. In a single blow, the size of the universe was doubled.[16] The measurement would go on growing until it reached what most experts believe is a pretty firm diameter of about 92 billion light-years.

Allan Sandage, another astronomer with whom Zwicky would feud, used the Hale Telescope to show that the universe is expanding in a uniform way in all directions, another proof for the idea that the universe came to life from a very hot, very dense material 13.8 billion years ago—the Big Bang.[17]

When Fritz Zwicky got his chance at the Hale, he would make up for lost ground, recording important discoveries relating to galaxy formation and clusters of galaxies, the largest structures in the universe. In scientific circles, he would build an entirely new reputation. No longer would he be the exploding star guy; now he became an expert on gigantic herds of galaxies. With George Abell, he would contend over the existence of superclusters,[18] which Abell believed in and Zwicky did not. He would also work on and catalogue dwarf systems, like the ones Hubble incorrectly attributed to Baade.

One of Zwicky's first achievements at the Hale was discovering vast amounts of material lying outside the structure of galaxies. At the time, the question over whether everything in the universe was neatly contained in galaxies or whether there was a lot of intergalactic material was a subject of energetic debate. Baade, according to Zwicky, didn't believe that anything worth thinking about would be found in the remote wildlands outside the galaxies. This would only fuel Zwicky's desire to prove him wrong. Zwicky believed there might be more matter between galaxies than there was contained in all the galaxies together, and in 1952 he produced a photograph that he said proved it.[19]

"Tempestuous, imaginative astronomer Fritz Zwicky has always been regarded warily by staid colleagues as a controversial man—part genius, part eccentric," noted *Life* magazine. "This week, the 54-year-old professor of Astrophysics at Cal Tech announced a discovery and released a photograph that may revise the geography of the universe." The magazine published a photo, covering most of a page (taken at Mount Wilson), showing how important the discovery seemed at the time. The image was of a bright filament of stars streaming between two galaxies, IC 3481 and IC 3483.

According to Zwicky, these filaments could be as much as "400,000 million million miles long." If such "bridges" prove to be the rule, "they may be found to contain so much matter that previous calculations of cosmic gravitational forces will need reworking." Zwicky's space bridges led to a lot of research in the sixties into strange galactic formations—tails, horns, and other odd deformities[20]—that may be caused by tidal forces between interacting star systems on a colossal scale. Like waves of water crashing into each other, these interactions occur over a time period not of minutes but of millions of years. Zwicky's space bridges never caught on with other astronomers the way some of his other ideas did, though some modern models of galaxy formation do show features where sheets and filaments intersect.[21]

Zwicky was nonetheless prescient in suggesting that space was not the same in all directions but could be quite lumpy in places. "This was in contradiction to mainstream thinking, which held that everything was homogeneous," said Robert Kirshner, the Clowes Professor of Science at Harvard University.

After announcing his discovery, Zwicky joked to a fellow Swiss that it seemed to be the job of their compatriots to "find the dirt in the universe."

Just weeks before Hubble got down to serious work at the Hale Tele-scope, on September 21, 1949, a small ceremony was held at the Los Angeles Air Force Base. Undersecretary of the Navy Dan Kimball pre-sented Zwicky with the Medal of Freedom after Air Force Brigadier General Thomas C. Chapman read the citation. The ceremony was followed by a luncheon at the Huntington Hotel.

President Harry S. Truman's decision to honor the Swiss physicist had been an easy one, really. Zwicky's groundbreaking work in jet propulsion, his research into the Nazi war machine, and his close interrogations of von Braun and other German rocket scientists were all valuable contri-butions not only to the war effort but also to the emergence of America as the world's leader in air power after the war. "As technical representative" of the US Air Force, "he contributed immeasurably to Air Technical In-telligence," the citation read.

Zwicky was the first non-American to win the award. The petition cited his "initiative, remarkable linguistic abilities, broad knowledge of physics and chemistry . . . together with an outstanding ability to exploit foreign technology in rockets, guided missiles and associated equipment for further utilization by the United States, [which] made his service most valuable to our war effort."

The language of the citation carried a strong hint as to how it came about. After the war, Zwicky's old war buddy, Colonel Jack O'Mara, took a leadership role in Air Technical Intelligence. There were few people, in or out of the military, who were greater boosters of Zwicky's special talents. O'Mara shared Zwicky's belief that the educated classes in Amer-ica, Fritz's colleagues at college campuses, especially, failed to appre-ciate the threat posed by Soviet Communism. "I think we can both agree that by bad management of this countries [sic] resources including human resources it could easily lead to our downfall," O'Mara wrote to his friend.[22]

The Air Technical Intelligence Center (ATIC) played a major role in US efforts to gather intelligence about Soviet movements and strategy in the fifties. Its influence declined after the U-2 spy plane piloted by Francis Gary Powers was shot down and captured by the Soviets in 1960. The plane carried a lot of highly classified ATIC equipment, which ended up in Soviet hands. Afterward, the head of Air Force intelligence "ordered

Zwicky receives the Presidential Medal of Freedom from Undersecretary of the Navy Dan A. Kimball, at a small ceremony in Los Angeles on September 21, 1949. Next to Kimball is US Air Force General Thomas C. Chapman. The award, for Zwicky's work debriefing the Nazi scientists and recommending steps the United States should take to stay ahead of the rest of the world in aeronautics, was the first time the honor was bestowed on a foreigner.

ATIC to stop developing technical data collection devices," O'Mara wrote angrily to his friend.

Given Zwicky's ongoing work for the Navy, that service might have been expected to also be a booster. But he had made enemies in the highest reaches of the naval service. That enmity would cost him dearly in just a few years.

By the dawn of the fifties, Fritz Zwicky's reputation for volatility had also reached the ears of the popular press. *Life*'s characterization of the "tempestuous, imaginative astronomer" would stick with him. There would be no more descriptions of the "dreamy-eyed professor" with chalk in his pocket.

Tom Gehrels, the Dutch astronomer who was so awed by the sight of the Hale Telescope, and who would launch a comprehensive survey of asteroids as part of a long career at the University of Arizona, had several memorable encounters with the Swiss astrophysicist, both on and off the mountain. Gehrels had been active in the Dutch resistance in World War II and was already a strong-minded young man when he came to America, a trait he needed the day he knocked on Zwicky's door, hoping to talk about the future of spaceflight.

"The greenhorn foreigner didn't catch the amusement with which he was directed to go see Zwicky," Gehrels wrote later of the man he called an "imaginative maverick."[23] So he "innocently trundled off to Pasadena, found the house, rang the bell, and waited." The door barely cracked open and a female voice asked what he wanted. Margrit's English was still poor, so her shyness with a stranger was understandable. Gehrels said he wanted to talk to Professor Zwicky about spaceflight. The door shut, then opened again to a booming voice.

"Whaddya want?" Zwicky demanded. Gehrels explained his mission. The door swung open, and the big man appeared in full aspect. "But don't you know? It's secret," he said.

A few years later, Gehrels was no longer the fresh-faced youth; he was by then a respected astronomer himself, with time allotted on the big Schmidt to observe asteroids, when he appeared at the Monastery. "Zwicky must have seen me walking up; before I could get to the steps," Gehrels wrote later, "the door swung open and here the big man stood again, booming, 'Who are you?'"[24]

"He knew, of course, but this was part of the act." Gehrels, twenty-seven years the professor's junior, explained that he had come to share the night with the famous star researcher. Zwicky peremptorily dismissed him, turning away. "Something had to be done, quick," Gehrels realized. "The ex-soldier's voice came to the rescue, thundering, 'Wait a minute, Fritz,' and you never saw a giant turn around so quick."

Gehrels announced that he had been given time at the telescope and would arrive at ten that night to begin his work. Zwicky was furious, or at least gave a convincing demonstration of fury. He strutted to the phone and called Bowen in Pasadena. Zwicky yelled into the phone until Bowen asked to speak to Gehrels, to see if the young man would compromise on his demands. Gehrels dug in his heels and asserted his rights.

Zwicky took the phone again, and yelled a while longer, while Gehrels stood by, thinking it was going to be "a godawful time on the mountain." At last, Zwicky barked, "I won't talk to you any more, young man," and hung up the phone (Bowen was just ten months younger than Fritz; both were in their late fifties when this incident occurred). Then Zwicky turned "with a naughty grin on his face, saying, 'Now let's have dinner.' We were the best of friends the rest of his life, and, oh yes, the times of switching were exactly ten and two, with courtesy. He would have loathed me if I had yielded."

This incident revealed everything about Zwicky the man, in the fullness of middle age and still a formidable presence: the bullheadedness, the arrogance and mischievousness, and, finally, the remarkable warmth once you cracked the carapace of his blustery eccentricity.

His longtime assistant at Caltech, Paul Wild, a fellow Swiss whom Zwicky went out of his way to bring on board at Caltech, told a similar story of an encounter with Fritz, shortly after the war. "I had been working with him only a few weeks, [and] we got into some argument," Wild wrote, years later. "He finally grew angry and said: 'Now listen, Wild, I am here since 25 years, and have argued with most of my colleagues, and was always right in the end. Now you just have come from Switzerland; don't imagine that you already have a good chance to win.'" With that, Zwicky stalked off. He returned thirty minutes later, walking up and down in irritable silence, before quietly admitting, "Well, almost certainly one day somebody else will be right, of course; but not you, and not now."[25]

With that, having given due credit to the odds-makers, he was off again. As someone who worked closely with Zwicky for many years, Wild came to know his moods. For him, the man's humanity was never far in the background, even during his explosive tirades. But not everybody got to see all sides of Zwicky.

Zwicky's Halley Lecture, touting the morphological method, so impressed the Oxford dons that they engaged him to write a book outlining his unique approach to problem solving. The book, to be called *Morphological Astronomy,* was duly submitted to Clarendon Press, which outright rejected it, citing his ungentlemanly "attacks on other astronomers." Rather than accept the fact that he had allowed his temper too free a rein, Zwicky demanded the manuscript back, so that he could find a new publisher, which he did, leaving the attacks on the astronomical community intact. "False notions of scholarship and gentlemanliness must . . . not prevent us from pointing out the salient facts," Zwicky wrote in the Foreword.[26]

Among the prominent astronomers Zwicky now engaged were a younger generation, who were building on the discoveries made by men like Zwicky, Baade, and Hubble. Zwicky had trouble accepting their researches as homage. Instead, he still felt the need to compete, to see their work as derivative, or even as outright theft. Far less certain of his place in history than he should have been, he felt the need to remind people of the things he had done, boasting immodestly that his contributions in rocketry were as important as those of Goddard and Konstantin Tsiolkovsky, the Russian theorist whose early work influenced von Braun and the other Germans.

But if he could be paranoid, it was not without reason. His complaint— that a small in-club of astronomers prevented him from using the big telescopes on Mount Wilson and Palomar until they had found the most interesting objects—was a legitimate one. His first night at the controls of the 200-inch telescope on Palomar Mountain was three years after work officially began there. In his diary, he wrote that he got "excellent photographs of distant clusters, intergalactic matter and dwarf galaxies."[27] All were much on his mind at the time.

His relationships with his colleagues in science may have been fraying, but the public remained solidly in his corner. Zwicky had become such a cultural icon that he even began to appear in fiction. In March 1952, the Peabody Award–winning radio show, "Halls of Ivy," devoted an episode to Zwicky's ideas. Written by "Fibber McGee & Molly" creator Don Quinn, the show featured Ronald Colman as the president of a small Midwestern university, Ivy College.

The episode played with the idea of supernovae—the source of Zwicky's fame for most people—and his unusual name. "Toddy, you know how I love to have these little bits of startling information at my fingertips, but

I don't think this is one I can use at the Women's Club," Vicky, played by Elizabeth Patterson, says to the testy college board chairman, played by Herb Butterfield. "Now, come on, confess! You made it all up, didn't you? Including Dr. Flicky."

Zwicky's popularity with the public, combined with his frustrations with colleagues, only encouraged his penchant for making grand pronouncements. With air and water pollution being recognized as growing problems, Zwicky suggested carving pieces out of other planets and dragging them to the inner solar system, where they could be made habitable for humans escaping our damaged world.

Zwicky's utopian musings eventually caught the fancy of the *New Yorker's* droll columnist, E. B. White, who penned a predictably excellent, tongue-in-cheek meditation on the new scientific age.[28] "Today I've been reading a cheerful forecast for the coming century, prepared by some farsighted professors at the California Institute of Technology. . . . Man, it would appear, is standing at the gateway to a new era of civilization," he wrote. "Technology will be king. . . . The population of the Earth will increase and multiply, but that'll be no problem—the granite of the Earth's crust contains enough uranium and thorium to supply an abundance of power for everybody. If we just pound rock, we're sitting pretty."

White considered it a "splendid vision: technology is king, Jayne Mansfield the queen." Noting that Wormwood, White's nickname for his home in Maine, was "well supplied with rock," he wondered whether his next move should be "to extract this stuff, or can I leave my stones be?"

The only place for a nuclear reactor at Wormwood, he said, was the brooder house, but "I need the brooder house for my chicks." Expressing doubts that were beginning to be raised about nuclear power, White wrote, "I am not convinced that atomic energy, which is currently said to be man's best hope for a better life, is his best hope at all, or even a good bet." He would feel better about technological progress if man spent less time trying to outwit nature "instead of respecting her seniority. Almost every bulletin I receive from my county agent is full of wild schemes for boxing nature's ears and throwing dust in her eyes.

"Dr. Fritz Zwicky, astrophysicist, has examined the confused situation on this planet and his suggestion is that we create one hundred *new* planets.[29] Zwicky wants to scoop up portions of Neptune, Saturn and

Jupiter and graft them on to smaller planets, then change the orbits of these enlarged bodies to make their course around the sun roughly comparable to that of our Earth. This is a bold, plucky move, but I would prefer to wait until the inhabitants of *this* planet have learned to live in political units that are not secret societies and until the pens on writing desks are not chained to the counter. Here we are, busily preparing ourselves for a war already described as unthinkable, bombarding our bodies with gamma rays that everybody admits are a genetical hazard, spying on each other, rewarding people on quiz programs with a hundred thousand dollars for knowing how to spell 'cat,'[30] and Zwicky wants to make a hundred *new* worlds."

13

DOMESTIC LIFE

FRITZ ZWICKY HAD always taken pride in his physical prowess and robust health. He enjoyed demonstrating one-arm pushups in the Caltech cafeteria, which his critics snickered was unbecoming of a tweed-jacketed professor of astrophysics. But for Zwicky, physical mastery went hand in glove with intellectual achievement; it was one more way for the man of genius to prove his worth and merit. He told friends that he hoped to live in three centuries, which once seemed possible. His body, however, was proving not to be up to the challenge of his will. In March 1952, at 54 years of age, he suffered a heart attack while skiing on Mount Wilson. At the time, he was showing some younger colleagues how to manage the slopes. He was incapacitated for weeks afterward. "I am just back on my legs now," he wrote to a friend in June.

While Zwicky prided himself on staying in good shape, he had in recent years fallen victim to the usual pitfalls of successful people in middle age: too much good drink and rich food, and an exercise regimen that was erratic. His weight had ballooned, and he suffered from gout. The heart attack was a shocking wake-up call.

Zwicky's younger brother, Rudi, died of a blood disease at 52, about the same time Fritz suffered his heart attack. This, coupled with his mother's early death from tuberculosis and his own physical infirmities, made him a ready consumer of the new cures and dietary advice offered by the medical profession. Margrit, who by this time had developed her culinary skills, now had to learn to cook all over again, without eggs and butter and quite so much red meat. Zwicky gave up his pipe and began walking to Caltech each morning.

Marshall Cohen, a colleague at Caltech, recalled a memorable exchange with Zwicky some years later on one of his morning constitutionals. Cohen liked to ride his bicycle to work along California Boulevard, which fronted Caltech. One morning, he passed Zwicky churning toward the university with his long, loose-limbed strides. "Hey, how are all those God-damned faster than lights going?" Zwicky shouted, not slowing his pace.[1]

Cohen had caused a commotion with the discovery of something called superluminal motion. Essentially, it appeared that he and his associates had found evidence that some objects traveled faster than light, violating Einstein's proposition that the speed of light was the top end of velocity in our universe. Zwicky, who believed that fundamental constants like the speed of light could indeed change over time,[2] was excited by Cohen's idea.

In the end, Marshall Cohen's superluminal motion turned out not to challenge but to confirm special relativity.

Even after adopting a healthier lifestyle, Zwicky feared that illness and infirmity might well prevent him from enjoying his family's companionship in old age. He began referring darkly to his "Homeric" battles with health. On several occasions, he told Margrit and the girls that he now saw one of his most important jobs as preparing them to be a widow and orphans.

If he felt assailed, and he did, in his work and by declining health, laurels continued to accumulate for him, proving that his contributions had not, as he often feared, been forgotten. The Medal of Freedom had hardly had time to gather dust on his shelf when he was invited to address the *Landsgemeinde* celebration in Glarus. This is an annual event that draws the entire population to a modest square to discuss the affairs of the community and to pass new laws, if any are thought necessary. Any citizen had the right to propose a new regulation. The assembled multitude voted it up or down.

Glarus in 1951 remained much as it had been in Zwicky's youth, unassuming but determined, uncomplicated but proud. The town was thrilled to have one of its own, a celebrated scientist and explorer of the darkest continents in the universe, among them. It brought him satisfaction to recall that only a few years before, the nation's leaders had sneered at his suggestions for national defense. Now, the Swiss people gathered around him in the hundreds, hanging on his every word. A sartorial problem developed on the big day. Zwicky's suit was deemed unacceptable, and a better one was found for him. Except the pants were too long, forcing him to tug at them as he spoke.[3] Nonetheless, his speech, about the work going on at Palomar, went off very well. "Only the acoustics there is not too good, so part of my film . . . could not be understood," he wrote home afterward.[4]

Buoyed by the experience, he began to believe that his problem-solving process, morphology, would soon find a larger audience. He began referring to morphology as a philosopher's stone and suggested that the process would soon pass William James's test of new and revolutionary concepts: they are "first mocked," then said to be "valid but not useful," before finally reaching the point where "everybody claimed to have thought it up."

As the fifties wore on, work continued to flow in at Aerojet, which had become one of the nation's leading rocket and missile firms. As such, the company was flooded with job applicants, including ex-Nazi scientists who had been repatriated in America to help win the Cold War. J. S. Warfel, vice president of avionics, sent Zwicky a letter asking "whether there is any way in which you can check the reputations" of some applicants.[5] He did so, even loaning one of them, Paul Schmidt, money to buy a house. It was Schmidt's research on "pulse engines" that led to the development of the V-1 buzzbomb. Zwicky recommended Schmidt after being convinced that the talented engineer had never worked at Peenemünde and had nothing to do with the bombing of London during the war.

With all this going on, Zwicky's home life was a sanctuary, one that sustained him through many battles still to come. Two years after Margritli's birth in 1948 came Franziska, named for Fritz's mother. Finally, a third girl, Barbarina, joined the family. Of the three, the oldest girl was the more dutiful and reserved, bearing much of the weight of family expectations. Fran was different. She was an active and adventurous girl, the one most likely to buck her father's rules. She was not a rebellious child in the way people think of it these days, but she was the one most likely to insist that she be allowed to go her own way.

The Zwicky home in Pasadena, a short walk from Caltech. Fritz bought it in 1948, months after his marriage to Margarita Zürcher. After several years, they added a study, where Zwicky could work, his desk piled high with papers, or concentrate on his hobby, stamp collecting.

"What I hated, my father was the keeper of sweets," Fran said in the 2015 interview at her home in Mettmenstetten. "Holes in the teeth was his department."[6] After returning home from trick-or-treating on Halloween, the girls had to hand over their candy. Their father put it on a closet shelf. After dinner each night, he would get one piece down for each of them.

This was a policy carried over from his childhood in Bulgaria, when he and his siblings would receive small bits of chocolate at Christmas. While the others gobbled theirs, Fritz, showing his customary discipline, put his aside, allowing himself one bite each day.[7]

While most American children were being coaxed to brush their teeth at bedtime, the girls in the Zwicky household were taught to use floss, as well. Fran's way of rebelling was to use a piece over and over, to her father's exasperation. "Father said 'get rid of that,'" Fran recalled, laughing. "He'd say stop acting stupid and get a new one. I said, 'I'll wax it again with a candle.' Oh, I was a terrible child. I always talked back to my father." Zwicky's anxieties over dental hygiene at least partly stemmed from the fact that his younger brother's death was attributed to disease that had spread to the heart from his rotted teeth.

One of Fran's childhood escapades provided an anecdote her father recounted years later in his German language book, *Jeder Ein Genie (Everyone a Genius)*, which laid out his philosophy that humankind could be happy and fulfilled, and that dictators and mass killers like Mussolini and Hitler and Stalin could find no support for their toxic visions, if every person found his or her true calling in life.

"Margritli was doing cartwheels on the lawn. My father was clapping," Franziska said six decades after the incident. "I couldn't do one. To this day I can't do a cartwheel. I was so unhappy that I stood on the sidewalk and said, 'Watch what I can do.' I let myself fall backward and cracked my skull on the sidewalk."

Her injuries were minor, but to her father, this was a perfect example of how a failure to find an outlet for healthy self-expression can lead one down a destructive path. It was a heavy load to place on the shoulders of a three-year-old. But Fran didn't mind being asked to stand in for Stalin. "I don't even remember the incident today," she laughed.

Fran knew all about her father's reputation in America. And it pained her. Whatever sins her father did or did not commit, to her he was just dad. "I don't recall a single time when he was harsh with us," she said. She likes to tell a story about his habit of rescuing wounded birds. At the time, she and her sister Margritli were still young girls, living on Oakdale Street. There were avocado and peach and grapefruit trees in the back yard, a swing for the girls, and an incinerator that the family stopped using when the smog got bad in the late fifties. "One time a pigeon fell into the chimney. Our father tried to smoke him out with fire. When we finally got him out, his wings were singed. We called him Willy. When he got his feathers back, my father took him outside and taught him to fly. He would pitch him into the air, over and over again. Every day he picked him up and tossed him into the air, until Willy got the hang of it. He would fly off during the day but then he'd come home at night. We had a tame bird. The next year the other birds came through and Willy joined them and left."

Over the years, she said, her father nursed several foster birds with the same concern. She looked up, with a question on her face: Do you see? How could such a person be as difficult as they say?

The age difference between their parents never seemed odd to their daughters. "I never noticed that my mom was super younger than my father," Fran said. She was mortified when a stranger thought Fritz was her grandfather.

Their parents didn't betray discomfort over the issue. "I thought they were super together. He really loved my mom," Fran said. Over the years, they developed a private communication system that "you could see and feel when they were in the room together."

"In the 60s, people called it called it good vibrations," agreed Margritli, the older sister. "You could feel the energy between them."

Her mother, Fran said, was "a cute, curly-haired blond and also kind of the tomboy" in her family. A memory she treasures was hearing her mother singing in the kitchen while she cooked, even yodeling, Swiss mountaineer-style.

Fritz was clearly the head of the household. It was the way everyone lived, then, even if the man of the house was not a famous scientist. Despite her youth and lack of sophistication, Margrit was not intimidated by her husband or his reputation. "My mom had a lot of gumption. I don't think he dominated her at all," Fran said.

Though her father had been a mountain climber and skier, by his sixth decade, when his daughters came along, he had shed his outdoorsiness for the coat and tie and other trappings of the mid-century college professor. His daughters never saw him in shorts. "He was born an adult," is the way Fran described him.

His manner of dress did change over time, however, a nod to the relaxing of codes of fashion in the sixties, as well as his experiences in the Southwest, at White Sands. He eventually shed the conventional four-in-hand for a bolo tie with a turquoise thunderbird clasp. According to another colleague at Caltech, Gerald Wasserburg, Zwicky sometimes even said he was native American, a ludicrous claim coming from a man whose Swiss-German accent was still very much in evidence. "He claimed all sorts of crazy things," Wasserburg recalled, smiling at the memory. "He was an extravaganza."[8]

As for the many stories of his outrageous behavior, the only shocking incident Fran could recall was the time he climbed up on the dining room table to demonstrate a principle of physics. "Mother didn't think that was so cool."

Margrit couldn't always control his outbursts, but that didn't mean she didn't try. When she chastised him, she would bark out, "'Fritz,' very sharply." If that didn't do the trick, she warned that if he didn't stop, she would leave. Sometimes, she did, getting in the Dodge sedan the family owned in the fifties and driving off. "She did that with us also," Fran said. "When we were little and acting stupid, she'd say, 'stop it.' If we kept on, she wouldn't say a word, but she'd go out to the car, and leave. We were scared she'd never come back so we would sit quietly and wait for her. I think she just drove around the block."

Though he enjoyed the occasional Old-Fashioned cocktail when he took his wife to dinner at the Athenaeum, the girls never saw their father in a state of diminished capacity. As for the marital shenanigans that went on in some households when the era of sexual experimentation arrived in the sixties, there was no hint of that.

Just months after the marriage, Rösli Streiff resurfaced, suggesting in a letter that her old friend pay her a visit on his next trip to Switzerland. By this time, the famous scientist was traveling a lot, giving paid speeches and appearing at international symposia. If Streiff was attempting to break up the new marriage before it settled in, she failed utterly. Zwicky replied with a sharp reminder of her earlier note, in which she chastised him for plucking an innocent still in her teens, and predicted the marriage would never last. He turned her down flat. "Despite my having little knowledge in such matters, it seems quite clear a woman does not like it when her man is invited alone,"[9] he wrote.

Streiff's position was at least understandable. As recently as late 1946, only months before he met Margrit, he sent Rösli a letter concluding with the words, "Greetings to all our friends and from far away a kiss for you." Rösli could hardly be blamed for developing the feeling that she would at last have the object of her heart. He had also adopted the more intimate German term, "du," in his letters to her. Her disappointment on discovering his marriage isn't hard to imagine, or justify. Though their communications continued afterward and even resumed their former friendliness, there were no more invitations. Rösli Streiff never married.

Zwicky's young family soon outgrew the confines of the small bungalow on Oakdale Street. In 1951, the couple paid $3,800 to add a study onto the back of the house, where he could retreat to work and find refuge from the running feet and squealing of children. One thing he especially hated was whistling, which is considered bad luck in some cultures.

"We liked to whistle," Fran recalled. "Especially the theme from the movie, The 'Bridge on the River Kwai.' He would yell out, 'Stop whistling.'" His hatred of whistlers was well known in the halls of the Robinson building at Caltech. Anyone whistling past his office would receive a thunderous rebuke.

Like his father, Zwicky collected stamps, and he enjoyed nothing more than retreating to his study with his stamp collection on a stifling Pasadena afternoon. "If we would run into his office he would say, 'Don't come in,'" Fran recalled. "Then we knew the stamps were out."

Journalists trying to win favor with the famous astrophysicist learned to gather stamps for him on foreign travels. Perhaps embarrassed to admit that he enjoyed such a noncontemplative hobby, he told the reporters that the stamps were for his daughters.

His desk was sacrosanct. "Nobody was allowed to touch his desk," Fran said, recalling it being piled high with papers. On the wall of the study, he hung a photograph of Mount Robson, in Canada, a cherished reminder of his climbing years. He had a lot of photo albums, too, which the girls liked to paw through when their father wasn't looking. In one photo, they saw him with a strange woman. She had long, beautiful, dark hair, and a self-confident expression. When Fran asked who she was, her father looked uncomfortable and told her to ask her mother. "That's Dorothy," Margrit said, simply. "They were married."

When the family went out to eat, Fritz preferred the Moser brothers' Swiss restaurant near the Los Angeles Coliseum. But he also liked to take the family out to eat at Eaton's, a restaurant at the Santa Anita racetrack, where he could get a steak for $2.30 and prime rib for $2.95.

The only problem, if they took the Fiat Fritz had purchased from a local dealer, was the crowding, with the girls elbowing for space and accusing each other of letting their petticoats invade the other's precious patch of seat. If Margrit drove, Fritz had a tendency to back-seat drive. If he complained too much, she would stop the car and get out, forcing him to take the wheel.

Fritz kept up his relationships with the "German boys," as Margrit referred to his group of ex-pat friends. Their meetings and meals began as once-a-month social gatherings, a kind of male coffee klatch that met to drink schnapps, debate politics, and fret over the declining morals of society as the unity and comity inspired by the war faded. Eventually, however, the get-togethers added wives and girlfriends, and they began mounting expeditions, sometimes to the Pantages Theater for a play or to the wine country in Napa. The meals grew from simple, plates-on-the-lap occasions to elaborate multicourse meals that had Margrit preparing for days in advance.

Despite the divorce from Dorothy, the friendship between Fritz and his former brother-in-law, Nick Roosevelt, remained strong. From time to time, the family took a road trip up the coast to Big Sur, staying with the Roosevelts at Point of Whales, overlooking the Pacific Ocean.

In the beginning, the girls attended public schools in Pasadena. Theirs was a typical childhood in many ways, though their father did have expectations when it came to their spare hours. Saturdays were spent learning French, a reflection of their father's belief that, as citizens of the world, they should be able to speak its major languages. They also had to learn to play bridge. "In America, people play bridge," Zwicky explained.

With the arrival of the sixties—the Beatles and drugs and hippies and the rest—the changes in American society left Fritz feeling increasingly alienated. Convinced that the public schools in Pasadena were not challenging enough, he enrolled Margritli and Fran at Polytechnic, an expensive private day school. Today, the tuition ranges from over $26,000 per student in the lower school to around $35,000 in the upper school.

That satisfied him no better. In some ways, things were worse. The discipline might have been better, but the rote memorization he considered "encyclopedic garbage," Margritli recalled.

"So what if you know the names of all the Presidents?" he fumed. Another thing he didn't like was that, while church and state were supposed to be separated, in practice, the Bible all too often found its way into the classroom. When he discovered that one of Margritli's teachers was reading verses in class, "That got my father. He complained to the principal and that stopped that," she said.

When Margritli was born, a newspaper reporter asked Zwicky if he planned to train his daughter as a morphologist. No, he did not want to choose her path for her. As a free and independent agent, she would make her own way, Zwicky replied. But what he saw as the waywardness of schools in America convinced him he needed to intervene, after all. He finally pulled the girls out of school in southern California and enrolled them in the Hochalpines Töchterinstitut Fetan (today, it is Hochalpines Institut Ftan) in Switzerland, an exclusive girls' school almost 5,600 feet up in the Alps. It was founded a century earlier, originally as a retreat for girls suffering from tuberculosis. The two-winged building, with an auditorium in the middle, housed eighty girls, many from the world's richest families.

"I was twelve when I was sent to boarding school," said Fran, who earns her living teaching English and lifeguarding. "Greedie (her nickname for Margritli) went to first year of high school" in Pasadena, before her father got fed up. "For him, it was a natural decision to send us, because he was sent away to school," Fran said. It wasn't a unilateral decision. "Mom agreed to let us leave."

For the two older girls, it was a bracing experience, but one they ended up feeling was good for them. "A lot of people thought our parents were evil because they put us away," Fran said. But she and Margritli were never resentful. They thought the experience made them strong and independent.

The school operated with Swiss-German efficiency. "We got up at seven to go to the auditorium, where we shake hands with the director and the director's wife, then the nurse, and we nod at the teachers," Fran recalled. "Then we sang a morning song. Everybody sang, every morning. Thanks to that, I know oodles of songs. There were different songs every morning. Then we go to breakfast. It was very formal, with twenty girls at each table, ten on each side, and a teacher at both ends. We had to use correct table manners. I was scared of the German teacher, Mrs. Hartmann. She was so strict."

It was all very regimented. But there was also time for recreation, and hardly a lovelier place for it could be found. In the summer, the girls went swimming in a pond near the campus. Above the school was a thick stand of forest, where the girls could go off hiking. In the winter, they skied on the mountain and ice-skated on the tennis court.

Among the girls, a strict hierarchy was observed. The older students were guardians and guides, called *schützpatrons,* for the younger girls, who were called *schützlinge.* The *schützpatrons* helped the young girls with homework. But they also functioned as substitute parents for the little ones, who were often scared and disoriented at being separated from their parents for the first time.

"When you turn sixteen, you become a *schützpatron,*" Fran recalled. "I had three or four girls under me. We lived according to age. The little girls were on the first floor, the middle ones on the second and *patrons* on the third. On Sundays, four of us *(schützpatrons)* would take twelve *schützlinge* and go mushroom hunting. We learned which ones were edible. We cooked them in butter at a campfire. We also went blueberry hunting. We would buy cream and pour it over them."

These excursions were fun, because there were no teachers to correct or challenge them. There was a small town in the valley below, called Scuol, but the girls were not allowed to go there or anywhere that they might encounter the male gender. "If you left the school there was a book you signed to let them know where you were going," Fran said. When some girls broke the rules and visited the town, it was a big scandal.

Fran liked to go alone for walks in the early evenings along a path called the *Philosopheweg,* or philosopher's walk, which snaked along a small creek. After dinner, the little girls were in bed by eight. "*Schützpatrons* would say good night to the *schützlinge.* We would sit on their beds and talk about the day, say good night and go on to the next girl."

This was a time for the older girl to be both parent and friend. "If a *schützling* was misbehaving, the teachers would call in the *schützpatron* and say they were acting badly," Fran said. Then it was up to the older girl to find a way to rein in her *schützling.* The rigor of the school and the separation from family could be too much. Some *schützlinge* washed out.

By nine at night, everyone was in bed, Fran often with the Latin she was straining to learn. Unlike her father, she had trouble with languages.

Finally, all went quiet, except for the swish of the alpine winds outside. Until the bell rang the next morning to "wham you out of bed," Fran said. She spent six years at the institute.

The girls were expected to write home every week, on Sunday. Over time, they began to lose their familiarity with English. Fritz took it upon himself to keep them up to speed. "Father had this obnoxious habit," Fran recalled. "He would copy my letters and correct them and send them back." In one note, he listed a "few words which you spelled incorrectly," including Siberia and coyote.[10]

He knew he was intruding, but he couldn't help it. In one letter, he asked her forbearance if "I harp around on" some mistakes but "it is important." In another, he asked her to excuse him "for sending so much advice."[11] In yet another letter, he criticized her for writing "alright" as one word. "I have never seen it written that way in the literature yet, or is that the generation gap?" he asked, referring to a subject that was much in the news at the time, as the baby boomers challenged the World War II generation's codes of conduct.[12]

Famous scientists and powerful administrators might have been intimidated by the magmatic eruptions of Fritz Zwicky, but his second daughter was unafraid. She told him his criticism was blocking her creativity. That put a stop to his meddling—mostly.

Of course, Fritz could not help being Fritz, and Fran saw him display his legendary brusqueness with outsiders during a family trip to Padua, Italy, for a conference. After a big, fine Italian dinner of scallopini and vegetables and potatoes, the waiter presented the bill. "Father blew his stack. He said he didn't order" the vegetables, and he wasn't going to pay for them. "Mom kept saying, 'Fritz.' The whole restaurant was watching. She said, 'If you don't stop we're going to leave.' He said, 'Okay.' We stomped out."

Margritli recalled the event as well. "Father said he didn't order it. The waiter said, 'but you ate it.' It was a horrific argument. Neither would give in."

Fetan, like Polytechnic, was expensive on a college professor's salary, even one with substantial earnings from outside contract work. Zwicky's diary for 1963 showed expenses of 9,000 francs per semester for the girls' schooling. Even though a Swiss franc was then only worth about twenty-five cents, that was a lot of money. "I don't know how he managed," Fran said. "He never talked about money to us. He drove the smallest car. They always flew economy."

At least until Fran became an airline hostess with Swissair and could get her parents better accommodations. Then they sometimes flew first class. They never flew together. This was another of Fritz's attempts to thwart the grinding of the wheels of fate. If one of their planes crashed, he reasoned, at least one parent would survive.

ON THE TRAIL OF ZWICKY'S GHOST

Did Dark Matter Rattle Its Chains?

A lot of physicists believed at one time that when the dark matter particle was found, it would be at the bottom of the Soudan mine. Some even thought the experiment has already snared the elusive particle.

Maybe, the experts thought, instead of fooling around with neutrinos, they needed to go big. The idea that the dark matter particle could be more massive than anyone expected fit with a popular theory, known as supersymmetry. According to that theory, each of the known particles—quarks, leptons, gluons, bosons, and so forth—should have a giant twin, known as a superpartner.

"We had a good understanding (of the particle universe) up to 100 proton masses," said Sunil Golwala, the Caltech physicist working on the Cryogenic Dark Matter Search (CDMS) experiment. "But we know physics is incomplete."[1]

It was suspected that the dark matter particle hadn't been found because science was simply looking in the wrong place. Out of this thinking came the idea that dark matter might be made up of what are known as weakly interacting massive particles, or, in one of the whimsical acronyms scientists are fond of, WIMPs.

"People began looking for the WIMP Miracle," Golwala said. The CDMS was one of the most ambitious efforts to find that miracle. Another experiment that has gotten a lot of attention, called DAMA (now called DAMA/LIBRA, short for DArk MAtter, Large sodium Iodide Bulk for RAre Processes), is located in another underground lab, this one under the Gran Sasso Massif mountain range east of Rome. That experiment has claimed since 1998 that it has detected dark matter.[2] But many other experts argue that if their detection is real, experiments elsewhere should have found it also. They have not. Several other experiments are in the offing, using the same sodium iodide detectors as DAMA, to verify the result or discard it, once and for all.

With CDMS, which is actually CDMS II, since there was an earlier machine at Stanford, the emphasis was on cold.[3] The heart of the device was one of the coldest places in the universe. At −459.6 degrees Fahrenheit, it was colder than outer space, a mere eyeblink above absolute zero, where everything grinds to a frigid halt. There, even the ceaseless bubbling of the quantum foam that is thought to be at the very bottom of everything that happens in the universe slows down.

The apparatus at the bottom of the mine was about eight feet tall and the width of a small meat locker. Much of that was shielding, twelve tons of lead and six tons of plastic, designed to keep unwanted particles from straying in to have a look around. The innermost layer of the lead is very special. Because recently manufactured lead is slightly radioactive, Golwala's collaborators went searching for very ancient lead, which has lost its radiation. The best ancient lead comes from Roman shipwrecks and the roofs of churches built in the Middle Ages. This fact has occasionally pitted physicists against archeologists and museums in competition for old lead ingots. Since the half-life of lead-210 is twenty-two years, even lead from a 200-year-old French shipwreck would reduce the radioactivity by a thousand times.

The elusive nature of dark matter requires the most extraordinary efforts to isolate it. The detector was put a half-mile underground to prevent cosmic rays from breaching the detector's defenses. But other particles, such as natural radioactivity from Earth, can still attack. To guard against this, the detector was placed inside a class-10,000 clean room that is thousands of times freer of microbes and dust than the air we breathe. Visitors donned special clothing and masks before entering. The room was also a Faraday cage that shooed away stray electromagnetic radiation. Jerry Meier, the jack-of-all-trades who watched over the experiment, periodically checked the room with a homemade electromagnetic detector consisting of a red ping-pong paddle wrapped with coils of wire. You don't spend decades working around physicists without learning a thing or two.

Getting the temperature down to bone-shattering levels inside the detector required a multi-stage process. First, the detector was cooled with liquid nitrogen and liquid helium isotopes. That reduced the temperature to −452 degrees. To get the extra bit of cold, forty-thousandths of a degree above absolute zero, a mixture of two isotopes of helium, helium-3 and helium-4, was circulated through the apparatus. These gases liquefy, forming two phases, one that is dilute in helium-3 and one concentrated. An interface forms between the two phases, much like the interface between water and air. Helium-3 evaporates from the concentrated to dilute phase, just like water molecules evaporate from our bodies while sweating into the surrounding air.

In the early days, according to Meier, the lab was using three or four big tanks of helium every week, at a cost of $3,400 each. At a time when there is a worldwide shortage of helium, that was a big problem.[4] Meier and the scientists at Fermilab solved that by adding a re-liquefier to trap the helium as it boiled off.

Given all this effort, it was a bit of a surprise to learn that the business end of the experiment was five small towers, each containing six hockey puck-shaped discs of germanium. Like silicon, germanium is a semiconductor. But it is not much used in the computer industry. Instead, it is used heavily in nuclear nonproliferation work, radioactivity screening, and to some extent in medical imaging.

"We went to germanium because it's kind of heavy," said Anthony Villano. He was a 32-year-old postdoc at the University of Minnesota. Balding and whip-smart, he has the unaffected self-confidence of a born-and-bred Midwesterner. His specialty was particle physics and his job at Soudan, along with other young team members, was monitoring the health of the CDMS II experiment. As with many physics experiments, it was the graduate students and postdocs who did much of the grunt work, while the professors stood on the sidelines, cheering or badgering, as the case may be.

The reason the experiment used germanium is simple. In physics, if you want to find something heavy, the best thing to put in its path is something about the same size and weight. "Think of billiard balls," Villano said. "The most efficient transfer of energy is from billiard balls of the same mass." While other substances, such as silicon or xenon, could also be used, it was felt that if the dark matter particle truly was a WIMP, germanium had the best chance of finding it. It was also a good candidate on which to place the critical phonon sensors—essentially, tiny microphones—that were designed to detect incoming dark matter particles.

Atoms of germanium, like other solids, arrange themselves in a particular way, known as a lattice. Once the lattice has been quieted by the detector's refrigerator, anything that passes through will ever so slightly shake the lattice, creating an infinitesimal sound wave, which in turn causes a barely discernible temperature spike. That spike tells scientists the type of collision that produced it. If the theorists were right and the dark matter particle is about the size of a germanium atom, the shaking of the lattice would tell them.

The thing that most worried Villano and Golwala, as well as their collaborators, was a rogue neutron getting in the detector, which could make a collision looking very much like one with dark matter. Neutrons could invade through particle reactions in the walls of the cavern.

Guarding against this danger was what the phonon sensors were designed to do. Each of the hockey pucks had four tiny microphones that tracked where the incoming particle hit and how much energy was deposited. Rogue neutrons, about 10 percent of the time, will "scatter" in the detector, hitting more than once. Dark matter would never do that.

Identifying dark matter is a tedious, laborious process, examining many, many interactions on a submicroscopic scale. In this underground vault, where all motion has been chained and every possible intruder locked out, they watched for the dark matter sneak-thief to venture in, stumble into a germanium nucleus, and rattle the lattice. After years of fruitless watching and waiting, physicists were about to throw up their hands when they decided to look at some older data, taken from a run that used silicon discs rather than germanium. Two, possibly three, events popped out. The mass of the suspect particles was a surprise. It was much less than the 100-proton size they expected, about eight or nine protons.

⁕ 14 ⁕

McCARTHY AND SPUTNIK

WITH A NEW YEAR, 1954, came a new opportunity for Fritz Zwicky to spread his message about the wonders of the new scientific age, one that particularly suited the anti-Communist cold warrior. He was asked by the US State Department, through its broadcasting arm, the Voice of America, to give a talk about the achievements at Palomar Mountain for broadcast behind the Iron Curtain. In just five years, Palomar had come to dominate the science of the stars. There was nothing coincidental about the timing of Zwicky's talk in February, coming as it did just months after the death of the autocrat Joseph Stalin, which led to violent protests in East Germany. In what became known as the People's Uprising, more than a million workers took to the streets, fighting pitched battles with their Soviet occupiers, who crushed the revolt with tanks, killing scores.

Watching all this from afar, the State Department hoped to further inflame the situation by promoting the great things Western science was achieving, things that East Germans struggling to feed their families amid food shortages could only imagine. Zwicky was flattered by the opportunity and delighted by the chance to employ his Russian language skills, which he taught himself.

The talk helped spread his reputation beyond the West. Ironically, given his lifelong, and energetically expressed, hatred of Communism, as his reputation suffered in the West, he would find validation in the

Soviet Union, leading to invitations to visit at a time when few Westerners were allowed in.

Even more ironically, while the State Department and Voice of America used Zwicky to sow discord behind the Iron Curtain, the foreign service agencies were being attacked in Congress for being stooges of Russian socialism. The revelation that the Soviets had the Bomb, as well as the victory of Mao's armies in China, had spooked the public and given a platform to a 46-year-old renegade senator from Wisconsin, Joseph McCarthy. The son of an Irish immigrant who had adopted the nickname Tail-Gunner Joe to celebrate his, mostly imagined, exploits in World War II, McCarthy found widespread notoriety in 1950, when he accused the State Department of being insufficiently patriotic.

Four years later, even as Zwicky was boasting about the superiority of Western science, McCarthy attacked the Voice of America, charging that its broadcasts were soft-pedaling Communism. This came as a rude surprise to the rabidly anti-Communist Zwicky, who could not have imagined that he would become a high-profile victim of the Red Scare that was about to sweep across Eisenhower's America.

McCarthy's attacks on the State Department and the Voice of America were more smoke than fire, but the damage they caused was real. One Voice of America engineer, his reputation ruined, committed suicide. That just whetted McCarthy's appetite. When the Army accused the senator of trying to blackmail the service into promoting a friend, he struck back, now charging that the military had been compromised. It was an audacious gamble, setting the stage for the famous Congressional hearings on Communist influence in the highest reaches of the defense establishment.

The hearings began on April 22, 1954, just weeks after Zwicky's Voice of America address. Televised, virtually gavel-to-gavel, over thirty-six days, they drew an audience estimated at 80 million, a significant fraction of the American population. The hearings revealed the power of television to educate and influence. Like the senator's previous investigations, the hearings produced little conclusive proof. But the event did provide great theater, and not the kind McCarthy desired, showing him to be a reckless bully, willing to smear private citizens on the flimsiest proof, or with no proof at all. The greatest drama, and most famous exchange, took place on day thirty, when McCarthy's assistant, the reptilian Roy Cohn, announced that the senator had a list of 130 subversives working at defense plants around the nation.

The Army's chief legal representative, Joseph Nye Welch, challenged Cohn to provide the names "before the sun goes down." At which point, McCarthy threw out the name of a friend of Welch's. Welch replied, "Until this moment, Senator, I think I never really gauged your cruelty or your recklessness." When McCarthy resumed his attack, Welch interrupted: "Let us not assassinate this lad further, Senator. You've done enough. Have you no sense of decency, sir, at long last? Have you left no sense of decency?"

That was the moment history would remember, the instant when McCarthy was finally pulled from his perch as the defender of American values. McCarthy might have lost his influence—he died three years later, likely of alcoholism—but the combustible combination of suspicion and fear that he put a match to survived him. If anything, the Communist-hunting zeal only intensified, as Fritz Zwicky soon learned.

In August 1954, D. L. Armstrong, one of Aerojet's top chemists, sent a warning letter to Zwicky in Switzerland. "We expect you to return with many ideas," he wrote, before sounding a note of alarm. "We thought that with the conclusion of the McCarthy–Army hearings the country might get back to business, but it appears there is a group quite willing to postpone important national affairs in order to perpetuate the polemic centering around [sic] McCarthy."[1]

In fact, in another historic irony, the Communist witch-hunt had now been taken up by the military itself. Once the prey, the military now became the predator. Zwicky's old foe, J. Robert Oppenheimer, who headed the Los Alamos laboratory that built the atomic bomb, was an early target. He lost his security clearance after the government reclassified him as a risk. With Oppenheimer, the suspicion was a natural one, since he did not disguise his political leanings. But there was hardly a more dedicated anti-red among the professorial ranks in America than Zwicky.

Still, he was about to become a prominent victim of the Department of Defense's new regulation, 2-203. Zwicky took care to quote from it in one of his diaries: Any "immigrant alien" eligible for a security clearance, it said, must have "formally declared his intent to become a US citizen."[2]

Zwicky had received a top-secret clearance in 1948 without much fanfare. A year later, he quietly resigned as director of Research at Aerojet after the Navy complained about his citizenship status. This was just cosmetic, however, since he stayed on as chief consultant. Now, that was no longer enough. Never mind that he had been instrumental to the war effort and was hard at work on defense projects that would protect America against future enemies, which most prominently included the Communist bloc. His hydrojets were designed to make submarines more effective, and rocket fuels derived from his and others' research in the forties would propel missiles farther and faster. Despite all this, in the atmosphere of doubt and suspicion prevailing in Washington, Zwicky was informed that his security clearance had been revoked, "since I do not want to become a citizen."[3]

Over the years, Zwicky gave a variety of reasons for his decision to remain a Swiss citizen. Shortly after his clearance was revoked, he said naturalized citizens in America were "second class" because their citizenship was subject to revocation under certain circumstances. It goes without saying that second class was not a status Fritz Zwicky would accept. On other occasions, he said he could fight Communism more effectively if he remained a Swiss citizen, since he could travel the world with fewer restrictions than an American. The popularity of America abroad, following the end of the war, was eroding by the mid-fifties, as fears of its hegemony grew. Von Kármán claimed Zwicky told him that he might want to become president of Switzerland someday. Whether Zwicky actually believed he could win that office is unknown, but the aspiration was not beyond his imagination.

Whatever the reason, Zwicky would not give in. When word of his fall got out, it provoked a political storm. "Step right up, folks, and watch a fascinating sight: Uncle Sam cutting off his whiskers to spite his face," wrote the *Democratic Digest,* a house organ of the Democratic Party, which was out of power during the Eisenhower years.[4]

"We now introduce Dr. Fritz Zwicky, one of the world's leading experts on rockets and jet fuels. . . . Because Dr. Zwicky has insisted on remaining a Swiss national, the Navy recently announced it had withdrawn his clearance for classified work." The Navy did not think he was a Communist, or that he was dangerous, or that he wasn't important to the national defense, the *Digest* said. Then why? Because of what it says "here in the iron-bound rulebook."

Joseph Alsop, author of a much-trusted political opinion column, called Zwicky three days after his clearance was revoked, proof of Alsop's superior sourcing. On June 26, 1955, a column titled "The Zwicky Case" appeared in newspapers across the country, citing the professor's service in the war and criticizing the military's short-sightedness. Drew Pearson, another respected columnist, quoted Zwicky saying that he could go to Washington "and give them my ideas—which I do. But I can't stay to see if they are properly carried out."[5]

"Zwicky, his head packed with vital secrets, could have no further contact with the classified projects that he has been supervising," wrote *Time* magazine.[6]

Zwicky's colleagues at Caltech went public with their outrage. In a letter to Charles E. Wilson, secretary of Defense, several, including von Kármán, noted Zwicky's "long record of devotion to the cause of freedom and of hatred for all totalitarian ideals whether communist or fascist." Withdrawing his top-secret clearance was "most discouraging to others who are giving their time and efforts to our national defense."[7]

Aerojet, desperate to retain its ungovernable visionary, launched a campaign of rehabilitation. In a letter to Thomas S. Gates, secretary of the Navy, Aerojet's president, Dan A. Kimball, himself a former secretary of the Navy, argued that Zwicky "has provided the original scientific thinking that has formed the basis of many presently active Defense Department projects. For instance, he could be termed the father of fragment chemistry, and has been very instrumental in progress made in the fields of boron chemistry, high-energy fuels, hydrojet engines, early warning systems, meta-chemistry (and) explosives."[8]

Currently, he was at work on "a new means of long-range detection which, under the code name of 'Rosy Glow,' will be classified secret." Which meant that the idea's author would not be able to work on his newest project.

Rosy Glow was aimed at countering a Soviet sneak attack. It was especially important to the military at the time, as fears of a surprise attack with nuclear-armed missiles were much on the minds of decision-makers in Congress and in the Pentagon. Soviet Russia's rapid subjugation of

Eastern Europe and its use of military force to quell uprisings in East Germany and Hungary convinced many experts that such an attack was a real possibility. Zwicky's solution was an early warning system that would alert the United States to a missile launch virtually instantaneously.

In his proposal to the military, Zwicky said he had considered all possible means of detection before concluding that the best solution would be to track the "virtual object" produced by the real missile. A large-scale launch would disturb the atmosphere, causing "sound waves, shock waves, turbulent eddies, slip streams of air, aerial blobs, and other perturbations." A launching missile would thus interfere with light coming from the stars, or the moon, or even the sun. These scintillations could be seen with good telescopes or cameras, placed around the Soviet Union. As proof, he submitted images of star tracks disturbed by a passing American jet fighter at Palomar Mountain's 48-inch Schmidt telescope.

"The obvious advantage of this procedure lies in the fact that the virtual object is enormously larger than the real object and can be therefore seen in principle although the real object may be obscured by intervening mountains, by clouds or, because of its great distance, by the Earth itself."[9] This was Zwicky's morphological method at its purest, ignoring the real object—the missile—and instead focusing on the way it disturbed the air around it.

While long-distance radar could detect incoming missiles only when they were well on their way, Zwicky said sensitive telescopes strategically placed around the Soviet perimeter could detect a sneak attack only moments after it began.

The military liked the idea so much it had given Aerojet a contract to develop it. Now, it appeared Zwicky would not be able to work on it. In fact, when he went back to work at Aerojet for the first time since the controversy erupted, in July 1955, he found his desk locked. A guard accompanied him to the restroom to make sure he did not take any secret information in there that he might try to smuggle out. It was the kind of unworkable solution that only a bureaucrat could come up with, since the intimate details of Rosy Glow were not on paper but in the professor's head, which tended to go wherever the rest of him went, guard or no guard. It was also an impossible situation for Aerojet, and it could not last.

More than a year later, with the problem still unresolved, Bill Gore, a former jato test pilot who now worked at Aerojet, told Zwicky the inside story of what happened. Gore recounted a conversation with an Air Force general "who seems pretty incensed that I do not take out citizenship," Zwicky wrote. But, according to Gore, it was "two or three men high up in the Navy who are holding up the clearance." The same men had blocked Dan Kimball, when he was secretary of the Navy, from giving Zwicky an award for his wartime service. It was the Air Force people, O'Mara and his colleagues, who got him the Medal of Freedom.

According to Gore, in Zwicky's diary entry, "there are only a few men who can do what I can and Aerojet needs me badly." Gore's openness concealed his real motive: He hoped that by letting down his hair to Zwicky that he could convince him to change his mind. If he would only compromise, Gore promised, he would be made "completely secure," with a big salary increase, to \$25,000 a year, "and an excellent pension. Actually without any necessity of ever having to retire."[10]

It was an impressive offer, from a man Zwicky considered a friend. Gore, a Marine Corps captain in the war, had been one of the first to test Aerojet's new jatos, when it seemed as likely the little rockets would tear the wings off the plane as that they would speedily launch the plane into the air.

But he was not dealing with a man susceptible to being cajoled, let alone bribed. If anything, the offer of a financial sinecure that would have helped him with the expense of giving his girls an exclusive education in the land of edelweiss and veal sausage would have simply caused the Swiss scientist to dig in his heels. Money was important to him, something he proved by refusing to give up his Caltech salary during the war, when there were few students. But he disdained those who sacrificed their integrity for a paycheck and would never allow anyone to think he might do the same thing. Especially once he had taken a public stand and won widespread sympathy for it.

Zwicky must surely have known that he was risking more than a government contract. He had to realize that he could not win a power struggle with Washington's faceless bureaucrats. His enemies in the Defense Department had more powerful weapons that they could train, not just on him but on Aerojet itself. The company had been a much fawned-over war baby, helping the Navy and Army Air Force carry the war deep into the Pacific. But with the missile race with the Soviets fully under way, it was now just one of many greedy nephews begging their uncle for the right to threaten the world with atomic ruin.

Aerojet had a lot more to lose than the cars and homes of its founders. The modest little rocket company had become an industrial and defense behemoth. From just $2 million in sales in 1946, its earnings a decade later, when McCarthy was toppled, had increased to $140 million. Three years after that, as it received contracts for the Titan, Polaris, Nike, and Minuteman nuclear missiles, earnings doubled again, to $300 million.

The little Azusa assembly plant had grown from twenty-five employees to 215. Kimball was already looking ahead. In 1956, ground was broken on a 20,000-acre facility outside Sacramento, where there would be no homeowners to complain about explosions in the night or clouds of smoke drifting over the new suburbs growing up around Los Angeles. Infected by Zwicky's enthusiasm, Kimball predicted that within ten years, "commercial planes will move along at speeds of 2,500 miles per hour."[11] The company was feeling so flush that it donated $400,000 for a new library at Caltech, to thank von Kármán for his vital role in the company's development. There were no similar gifts for Zwicky, though Kimball did tell him, in the midst of his security imbroglio, that "with my ideas the company will live another 20 years."[12] Perhaps, but would he see that future?

Given all this, there was little doubt who would win a war of wills with America's defense establishment. At first, both sides tried to make the new arrangement work. Zwicky boarded the Super Chief to meet with top Air Force scientists on his detection devices. Eventually, however, the military concluded that star scintillation had a major weakness: clouds. How, they asked Zwicky, would his system work when the sky was overcast? Zwicky argued there were methods to deal with the problem, but the Air Force then suggested that simple wind gusts might cause scintillations similar to a missile launch. Would the United States be guilty of unleashing a nuclear holocaust when the frigid buran blew across Siberia?

Rosy Glow never got any farther than the investigation phase. The same thing happened later with another of Zwicky's innovative ideas. In December 1956, *Aviation Week* said the military was looking into the possibility of using free radicals in rocket fuels.[13] Free radicals are molecules with a missing electron. As such, they are highly reactive, searching like ghostly widowers for their missing partner, and willing to steal it from any available source. This chemical neediness was something Zwicky thought could be exploited to make more powerful fuels. "Although the existence of free radicals has been known since 1900, the first suggestion that controlled free radicals might have an application to rocket propulsion came in a proposal to the AF (Air Force) Office of Scientific Research from Dr. Fritz Zwicky," the magazine said.

Zwicky turned to his old friend, Tadeusz Reichstein, for advice on the idea. As it turned out, a Swiss firm was also interested in new propellants, which amused Zwicky. "This is a laugh," he wrote, noting that years before Swiss experts "had violently criticized my attempts to strengthen the Swiss military."[14] In the intervening years, "propellant chemistry based on my original work on light element hydrides and nitrides has now grown into a billion dollar affair."

The trouble with free radicals, however, was that they were so unstable that taming them, necessary for controlled burning in a rocket, was problematic.

It was obvious that Zwicky's usefulness to the military was coming to an end, but just then, another of his ideas that had been written off came back to life. In July 1955, only months after Zwicky lost his security clearance, the United States and the Soviet Union both announced plans to launch satellites during the International Geophysical Year (1957–1958).[15] In the scientific world, and among the small cadre of rocket enthusiasts, the announcement was exciting news. Zwicky was delighted. Finally, people were beginning to recognize that they might do something with rockets besides attach bombs to them.

For years, since the failure of Zwicky's 1946 "artificial meteors" launch, satellite had been a dirty word among both scientists and military planners. Few scientists thought there was anything worth exploring in the wasteland of space. Hadn't Baade and Zwicky and Hubble and the other great astronomers at Mount Wilson and Palomar revealed the cold cruelty of the territory beyond Earth, a graveyard of dying stars and parched planets? But suddenly, with the International Geophysical Year announcement, the world seemed to be enthusiastic about space.

"People start calling me up like mad," Zwicky noted in one of his diaries.[16] Aerojet received a contract to build the rocket's second stage, which it delivered in the spring of 1957. The American Rocket Society was holding its meeting at the same time, and Andy Haley implored his old friend to address the session, which was expected to draw a thousand rocket fans to Washington. When Zwicky demurred, Haley wrote back in wounded disappointment, "It doesn't seem right not to have the Great Zwicky present."[17]

Robert Goddard may have been the true father of rocketry in America, but over the previous decade, few had done more to promote space than Zwicky. Hollywood had been trying for years to win his help for films about UFOs. He was invited to see a special screening of the film, "The Day the Earth Stood Still." The movie featured a visitor named Klaatu, who bore a warning that powerful forces outside Earth were concerned that humankind's warlike instincts were a threat to the orderly operations of the universe.

Even earlier, in 1949, Zwicky was invited to watch the filming of Hollywood's first serious treatment of space travel, George Pal's "Destination Moon." Employing impressive special effects, for the time, the film described some of the real-life perils of a trip to the moon. Zwicky was not listed in the credits—the science fiction writer Robert Heinlein did much of the work on the script—but he was photographed on the lunar set, waving to the people of Earth. Or was he beckoning them?

Zwicky didn't record his reaction to the films, but his views on aliens were well known. While he was sympathetic to the theme that humans were in a mess of their own creation with the atomic bomb, he found the idea preposterous that some outside force would help put things right.

The plans to launch satellites galvanized the enthusiasts. An international consortium was formed to coordinate the so-called Moonwatch program, and teams of amateur sky-watchers readied themselves. The general public, however, took little notice. Most news outlets either gave the report of the planned Soviet and American launches a brief mention or ignored them altogether.

All that changed on October 4, 1957, when a beachball-sized device beeped its way across the sky, provoking fear, awe, and something close to panic in America.

That evening, representatives from a dozen nations involved in the satellite program were relaxing at the Soviet embassy in Washington, DC, when Richard W. Porter was called to the telephone. Porter was chair of the technical panel for the US launch program. He was also the man who, with Zwicky, first interviewed Wernher von Braun, hiding in a mountain ski resort as Nazism collapsed. When Porter returned to the group, he said, "Gentlemen, our Russian colleagues are to be congratulated. They have successfully launched an earth satellite."[18]

A few blocks away, at the Smithsonian Institution, teams of Moon-watchers sprang into furious activity, trying to track the 22.8-inch diameter, 184-pound satellite as it orbited Earth, its tiny, one-watt transmitter beeping every .3 seconds. Sputnik's transmission was finally picked up at 8:07 P.M., Eastern Standard Time, and the news went out to millions of Americans.

For Americans of a certain age, the appearance of Sputnik in the skies—and it was indeed possible to see it pass overhead every ninety-six minutes—was one of those events that became a benchmark for their journey through life. Like the Kennedy assassination, the moon landing, and the Cuban missile crisis, Sputnik not only marked a point of departure on the map of experience but also created a new topography. Suddenly, not even the heavens were safe. The Soviets were up there, looking down on us, even if the little satellite had no camera.

President Eisenhower and his men didn't initially appreciate the impact the satellite would have on the public. Lee DuBridge, president of Caltech following the retirement of Millikan, described the reaction in the American scientific community. "The Russians put one over on us—fairly and squarely," he wrote several months later, when the furor over Sputnik was at its peak. "OK, we will win the next round."[19]

Some leading scientists, echoing Porter's slap on the back, even found reason for optimism. "In his millennia of looking at the stars, man has never faced so exciting a challenge as the year 1957 has suddenly thrust upon him," wrote astronomer Fred Whipple.[20]

But as DuBridge noted, the mood was far different in American living rooms. "To those who had not previously thought much about Earth satellites the reaction was about as violent as though Russia had landed an atomic bomb on New York . . . disbelief, hysteria, anger, recriminations, disillusionment, fear, and a hundred other emotional responses." Abroad, the reaction was only slightly less hysterical. "A great idol had toppled, (the) proposition that America was always first," DuBridge said.

For the Eisenhower administration, the initial reaction was mostly irritation that the Russians had gone off on their own and launched their satellite without letting anyone know what they were up to. Even as discussions to coordinate the international launch effort were going on at their own embassy.

Eisenhower's unruffled response was a serious miscalculation, contributing to the perception that the old general was out of touch and fusty, a duffer shambling around the golf course as the world rattled with danger. The fact that his administration helped create the conditions that caused the public panic by portraying the Soviets as automatons waiting for the right moment to launch a nuclear sneak attack was entirely lost on Eisenhower's men, as they scrambled to explain what happened and why things were going to be okay.

DuBridge noted that people in the United States had talked about space travel for years, "but they were dismissed as starry-eyed visionaries or just plain nuts." He might well have been thinking of his colleague at Caltech, Zwicky, who had been called both.

Zwicky himself lost no time claiming that Sputnik had vindicated him. If American politicians and scientists had only listened to him in 1946, the Soviets would never have beaten us to space. If "Americans don't use their knowhow and give more support to space research, Russia is apt to win the (larger) race for space," he told the *Los Angeles Times*.[21]

The big problem, according to Zwicky, employing a typically colorful metaphor, was that America had "too many yes men supporting mediocre or downright worthless scientific projects. We have too many no men obstructing or impeding advanced projects. And we have too many people who don't care."[22]

Von Kármán linked the failure to America's quiz show mentality. "American education is based on knowledge, while the European school system is grounded in thinking," he said.[23] "This is exhibited in our television quiz shows, where great amounts of money are given to persons whose memory enables them to come up with specific facts. . . . Scientific developments, such as Earth satellites and interplanetary rockets, are the result of thinking." It wasn't that American youth were lacking in skills, he said. American boys—girls were so scarce in the hard sciences that he didn't even bother to use the ungendered word children—"have more mechanical talent than Europeans." They also had imagination, he said, thinking of his students at Caltech, the Kármán Circus, and the Suicide Club. But these assets were wasted in the soul-deadening rote-learning environment that prevailed in American education.

In December, two months after Sputnik's launch, the diminutive Hungarian aerodynamicist appeared as the featured guest on the respected TV news magazine, "Face the Nation." Von Kármán—his bushy eyebrows, Dr. Strangelove accent, and wry smile firmly in place—faced a deeply alarmed panel of journalists.

His impish equanimity troubled Charles Von Fremd of CBS News, who challenged, "You sound not too frightened." Fremd said he had just returned from the newly opened launch facility at Cape Canaveral, where scientists told him "we are five to ten years behind the Russians."

"I am not quite so desperately afraid," Kármán agreed.[24] He said he was only mildly surprised by the Soviet success. Their science had always been good, he said, but their manufacturing was deficient. That they could mount a successful launch on the first try was the shock, though he said it was unknown whether there had been other attempts that failed, and which they did not announce.

For the panel of journalists, the real threat posed by Sputnik I, and its successor a month later, Sputnik II, was the revelation that the Soviets had rockets capable of penetrating space. That meant they had rockets that could target the United States. Von Kármán said that was certainly true. The question was how accurate would such a missile be? Well, okay, the panel pressed. But even a near-miss would cause tremendous damage. "It would certainly kill many people," von Kármán agreed.

"So at this moment there is very little reason for optimism, is there not?"

Von Kármán replied that he was a scientist, not a policy expert, but he continued to believe there was no reason for hysteria. The interview concluded with more hand-wringing about the failures of American education, and here von Kármán was in sympathy with his interrogators. Asked whether it was time to reduce the emphasis on liberal arts and place more on math, physics, and chemistry, von Kármán agreed that the hard sciences, his specialty, deserved more attention in American schools.

Von Kármán may not have seen cause for panic, but he was in the minority. The day after his appearance, the *Pasadena Independent* newspaper screamed in a banner headline, "Surprise Red Missile Attack Could Knock Out NATO Retaliatory Power."

Von Kármán opposed crash programs. He preferred steady, deliberative labor toward a shared goal, which was how he worked. But a crash program was exactly what was initiated by the Kennedy administration, when the new young president declared that an American would set foot on the moon before the end of the decade of the sixties. That goal would be met, though the president would not live to see it.

In the meantime, decision-makers in Washington realized they had to do something, and quick, to meet the Soviet challenge. Work was speeded up on the American satellite, called Vanguard, but it would not be ready until December.

So, after an eleven-year wait, Zwicky finally got his second chance to launch his little moons, or, as he now called them, "artificial planet zero." After the war, Zwicky had befriended Maurice Dubin of the Air Force Geophysical Research Directorate. Zwicky quickly won him over, convincing him that shooting pellets into space was not only a worthy political goal to outmaneuver the Soviets but also had a scientific purpose. It would yield important clues about Earth's gravity at extreme altitudes as well as the chemical makeup of the ionosphere, he said.

As early as 1954, Dubin began looking for an opportunity to make room for Zwicky's package of space marbles. As fortune would have it, the chance presented itself in mid-October 1957, just days after Sputnik launched. Was the date merely coincidental? There was no proof, but the timing was certainly suspicious, given the nationwide trauma Sputnik caused. The rocket was one of Aerojet's Aerobees, a high-altitude sounding rocket that was made for the Navy, based on the German V-2 model. Standing just twenty-six feet tall and weighing 1,600 pounds, more than a thousand of them were built over the years. Its maximum altitude was about seventy-three miles. The liquid explosive used to eject the moonlets was an Aerojet product developed by Zwicky and his colleagues known commercially as Jet-X.[25]

40,000 MPH IN SPACE

U.S. Fires Meteors

BEDFORD, Mass. (INS) Air Force scientists announced yesterday they have fired the world's first artificial meteors into outer space at a speed of 40,000 miles an hour—the fastest speed ever attained by a man-made thing.

At least two of the meteors or "sky pills"—presumed to be approximately the size of hand grenades — blasted upward from over New Mexico last Oct. 16. They headed into the unknown at 55 times the speed of sound, in the general direction of the sun.

The final, propelling blast had an intensity of nearly -10 visual magnitude, more than 5,000 times brighter than the Soviet Sputnik.

Dr. Maurice Dubin, physicist and project scientist of the experiment, said mankind's first artificial meteors may now be some 200 million miles in outer space.

TO JUPITER

He added, "they could be half-way to the planet Jupiter, or headed for absorption by the sun."

Dr. Dubin announced results of the experiment at the Geophysics Research Directorate of the Air Force, Cambridge, Mass., research center near Boston.

The artificial meteorites, like real ones, are simply objects travelling through space. They have no fixed course, but could wander or come to rest somewhere in outer space, including on the moon or Mars.

However, Dr. Dubin noted that the sun dominates our space, and scientists believe the "sky pills" probably were destined to be attracted and absorbed by the sun.

Satellites, such as the Russian Sputniks—as contrasted from these meteors—have not escaped completely from the earth's gravity. It is the combination of the satellite's velocity and the earth's gravity that keeps them in a generally fixed orbit about the earth.

As many as eight meteors.

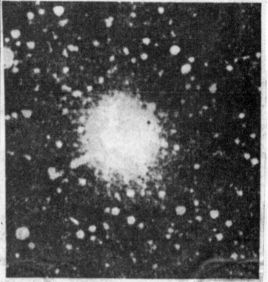

—United Press
Air Force camera follows trail of meteor, center.

Newspapers around the country gave prominent coverage to the successful launch on October 16, 1957, of Zwicky's bundle of metal slugs, which he called Artificial Planet Zero. This account appeared in Zwicky's hometown Pasadena *Independent.* For the rest of his life, Zwicky would claim, with some justification, that he was the first to put manmade objects into outer space, beyond Earth's orbit.

On October 16, in the midst of the worldwide hyperventilation surrounding the Soviet coup, Zwicky quietly traveled to Holloman Air Force Base in New Mexico, where he personally prepared the aluminum pellets to be inserted into the nose cone of the Aerobee. The pellets—the micro-moons, artificial planets, whatever—were to be ejected less than two minutes into the flight. Unlike in 1946, there was no fanfare surrounding this event. No journalists were on hand, either because the military wasn't sure the launch would succeed, or because it did not want to be seen trying to compete so directly with the Soviets, and with a rocket much less capable than the one that sent Sputnik aloft.

At 10:05 P.M., Mountain Standard Time, the Aerobee rose from the launch pad. Ninety-one seconds into the flight, a bright green flash was seen at Palomar Mountain, 600 miles away by Zwicky's assistant, Howard Gates, the younger brother of Zwicky's former wife. The flash was brighter than any star in the sky.

The pellets were shot out and away from Earth at an altitude of fifty-three miles above Earth. Measurements from the ground showed that they traveled at more than nine miles per second. Since the atmospheric pressure at that altitude was just a millionth of what it is at sea level, Zwicky concluded that the pellets possessed almost twice the kinetic energy necessary to escape Earth.

"Small man-made projectiles were launched away from the earth for the first time, never to return," Zwicky wrote triumphantly in his book *Morphology of Propulsive Power*.[26]

Anticipating critics who would argue that the launch was still nothing more than a stunt, Zwicky argued that this was simply the first stage in a process that "will pave the way for direct experimentation with all the planets and their satellites in the solar system." He dusted off his decade-old portrait of space travel, first outlined in his Halley lecture: "First throw a small bit of matter into interplanetary space, 2) then a little more, 3) then a shipload of instruments, 4) and then ourselves."[27]

The military did not announce the success at Alamogordo for more than a month, on November 22. But when it did, the press and public responded just as the Swiss scientist expected, with overheated proclamations that the United States had finally met the Soviet challenge.

"U.S. Meteors Shot into Space," proclaimed the *Los Angeles Examiner* in a headline covering half the front page, befitting a declaration of war.

By then, Sputnik II had launched, carrying the first living organism into orbit, a mongrel dog named Laika. Though the Soviets claimed the dog, a stray picked up on the streets of Moscow, lived several days in space, it was later shown that she died within hours of the launch due to over-heating in her container. Nonetheless, Sputnik II proved that, under the right conditions, life could survive in space, something that was very uncertain at the time. That discovery prepared the way for human spaceflight.

While Fritz Zwicky's humble "moonlets" could not compare to that achievement, he seized on the fact that the two Soviet satellites never reached interplanetary space, remaining in Earth's gravitational embrace until they burned up, weeks and months later. His meteors, he maintained through the rest of his life, "got away from the gravitational pull of the earth to become tiny satellites of the sun."[28]

Not all of them, apparently. A woman in Hillsborough, California, reported the day after the launch that strange bits of metal rained down on her house the night before. So while Dr. Zwicky's assertion that he was first to reach interplanetary space was up for debate, it could not be disputed that he was the first in history to produce space debris.

While many experts dismissed the launch, others saw a potential sinister purpose. *Time* magazine said the innocuous-seeming pellets might in fact be "satellite killers." "In spite of its speed a satellite is a sitting duck," *Time* said.[29] If any satellite "makes itself objectionable to a nation that it passes over, a defending rocket can be sent up to meet it." Zwicky himself never said or wrote anything that would indicate that he and his supporters in the Air Force were interested in militarizing space. He always maintained that his only interest was research. *Time* didn't name any sources for its speculation. But, as Zwicky's work in biowarfare and underwater propulsion showed, the Soviet threat had inspired a desperate need to counter the Soviets on the battlefield, under the sea, and even below ground, with the Terrajet. Space was simply the next arena of potential combat.

After his artificial planet zero launch, Zwicky took to the media, celebrating his achievement and striking out at critics who said it would never work and, even if it did, was useless for science. In a radio broadcast, he said that the failure to follow up on his 1946 attempt "allowed the Bolsheviks to cash in on a tremendous propaganda victory." If he had been allowed to continue his experiments eleven years earlier, the United States would have easily beaten the Soviets to space.[30]

This interview took place in February 1958, by which time the United States had finally managed to successfully launch its own satellite, Explorer I. Two earlier attempts to launch the Vanguard satellite had failed miserably, subjecting America's fledgling space program to another round of ridicule and inflaming the public once more. The press referred to Vanguard derisively as "Kaputnik."

"I never expected to live to see the day when a leaky fuel pipe would be regarded as an international tragedy," commented DuBridge. "Have they proven that a dictatorship is superior to a democracy?" Not in his mind. But others were less sure.

Wernher von Braun had predicted failure. He had a rocket ready, but the government, wary of his Nazi past, didn't want the first American achievement in space to be yet another proof of German technical superiority. With the successful launch of Explorer I, however, von Braun's role could not be hidden. With gritted teeth, the government allowed him to pose for a memorable picture at the Jet Propulsion Laboratory, the successor to von Kármán's Guggenheim Aeronautic Laboratory at Caltech. He was shown with Porter and JPL's director William Pickering, holding a model of the rocket overhead in celebration.

Zwicky dismissed Explorer as a pale copycat. "Because of the much greater weight of Sputnik I and II, neither the propaganda value nor the technical achievement of launching an American satellite, the Explorer, seemed very impressive," he wrote. In contrast, "only twelve days after Sputnik I, we shot for the first time in the history of man . . . artificial meteors into interplanetary space."[31]

He intended to "vigorously" follow up on that achievement. He announced plans to shoot his meteors at the moon and other planets. Analyzing the planetary debris caused by the impact would provide invaluable information about the surfaces of other solar system bodies. He was particularly keen on thumping the surface of the moon to see if there might be water hidden in the soils and rocks of the barren landscape. Zwicky argued, and wrote, that there was every reason to believe human beings could live quite comfortably on the moon.

He never got the chance. His role in the exploration of space, which had begun with experiments in jet propulsion in the midst of war thirteen years earlier, ended at Alamagordo. He would have nothing to do with the great achievements NASA would make in the sixties and beyond.

The space agency finally did get around to sending a battering ram at the moon to search for water. Zwicky's name was nowhere mentioned at the Ames Research Center in Mountain View, California, late one night in 2009, when a thousand or so people gathered to watch on a giant outdoor screen as the Lunar Crater Observation and Sensing Satellite—LCROSS—plunged into Cabeus Crater.

Combining showmanship and imagination, the event would have been one Fritz Zwicky would have enjoyed attending, or even sponsoring. Though he would of course have pointed out that it could have been done decades earlier if they had just listened to him.

And it did find evidence of water.

* 15 *

BLUE STARS AND QUASARS

BY THE TIME THE new president, John F. Kennedy, moved into the White House, in January 1961, Fritz Zwicky's security situation had become intolerable. The military refused to back down, and Fritz was as immovable as ever. Having taken a public position, he wouldn't consider changing his mind. He also thought his talents were so unique and irreplaceable that the company couldn't get along without him. He was wrong. Since the end of the war, Aerojet had attracted many talented specialists whose knowledge of the chemistry of rocket fuels and new kinds of metal alloys surpassed his. None, perhaps, could apply his broad insight to a problem, but the bulk of Aerojet's work, building America's nuclear-armed intercontinental ballistic missile fleet, didn't require that sort of thinking, anyway. Money was pouring in, and everyone was doing well. Tolerance for the company's outspoken house genius had reached its end. In late September, Bill Zisch, Aerojet's executive vice president, sent Zwicky a letter that, while measured and respectful, was unmistakable.

"I have been hopeful that I would hear of action you had initiated declaring your intent to become a citizen," Zisch wrote.[1] Since he hadn't, Zisch said he had reluctantly concluded that it would be "impractical to continue to be effective in your efforts for the company because of the highly classified nature of the work we are performing for the Department of Defense."

In as delicate a way as he could manage, Zisch was firing the man who had been so instrumental in the growth of the company. Zwicky waited a month before answering with what he said would be a few thoughts, but which consumed three pages of wounded pride and resentful assertions that Aerojet did not sufficiently appreciate him.

Since he was planning to retire in two years, when he would be 66, he said there was little reason, at such a late date, to apply for citizenship. By that time, he said, he would be spending only three months each year in Pasadena, anyway. The rest of the year, "I shall be roaming around and be occupied with activities intended to aid the survival of the free nations."[2] As self-aggrandizing as that sounds, Zwicky believed exactly what he said.

Given all this, he suggested that Aerojet simply wait for his retirement. He next cited his past achievements, which Zisch knew well. After the war ended, he added, many experts doubted there would be any use for rockets in peacetime. According to Zwicky, Aerojet executives even toyed with shutting the company down for good when the Germans and Japanese had been vanquished. Recalling his comment to the *Saturday Evening Post* that people would soon be traveling by rocket from Los Angeles to New York in less than twenty minutes, he groused, "It is too bad for the U.S. prestige in the Cold War that nobody backed me up then."

He then got personal, accusing Zisch of threatening him, saying that was something "no free man can accept."[3] He reminded Zisch of the time he considered quitting the company. "I tried to prevent that by giving you some friendly and, as I think, enlightening pep talks about the future of the company."

This was Fritz displaying his willingness to employ any and all weapons in a battle of wills, including guilt, shame, and accusations of ingratitude, heedless of the effect on the recipient. He might have considered himself fortunate that the company stayed with him over the previous six years since he lost his security clearance. Instead, he sought help from powerful allies, including Andrew Haley, the man who recruited Zwicky to Aerojet in the first place. Haley had moved on, opening a law practice in Washington, DC, specializing in the new field of space law.

"I certainly hope that the company you did more to create than any other scientist, will continue to reimburse you on an adequate basis,"

Haley wrote.[4] He added that he "would be happy indeed to help you protect your rights."

None of these efforts succeeded. On January 17, 1962, Zisch wrote that he regretted their association had ended, but there was "a price to be paid" for the fact that the work Aerojet was doing was "cloaked with security regulations."

As Zwicky left Aerojet in rancor and recriminations, his longtime friend and patron, Theodore von Kármán, traveled to Washington to receive the National Medal of Science from President Kennedy. It was the first ever awarded, and there had been many candidates. Von Kármán's selection reflected his contributions to aeronautics and, just as important, his unique talents in bringing diverse people together to achieve remarkable things. At 81, Zwicky's old friend was frail and walked on arthritic feet. His time in the limelight was all but over, but his intellect and wry humor were intact. At the Rose Garden ceremony at the White House, he swapped jokes and stories with the young president and his fashionable wife.

Zwicky, the occasional target of the older man's jibes, penned a letter to his "Todor."[5] "Now that you have reached another landmark in your great life it is exhilarating to reminisce on our pleasant and fruitful association during the past thirty-five years."

Zwicky thanked his old friend for going to bat for him and "scores" of others "whenever you thought we were right." Zwicky knew that without von Kármán's support, he would never have been chosen for the sensitive mission to Germany at the end of the war, where he served with such distinction. He then reminded von Kármán of the term "micro-Zwicky" that von Kármán had proposed as a measure of the roughness of a solid surface. "Hearing about your proposals I have again been impressed by your deep insight that the salvation of the world depends on sound construction as well as on a profound sense of humor."

Several months later, von Kármán was dead. From the estate, Zwicky received a stained-glass picture and two small vases, paying fifteen dollars. Zwicky also stepped in to settle a debt the estate owed Swiss bankers, turning over $25,000.[6] It is unclear whether this came from Zwicky's own funds. The esteem in which he held his old friend was such that it was possible, although von Kármán was certainly not destitute, owning a home on a better street in Pasadena than the one on which the Zwickys lived.

Fritz may have been out of one job—he maintained his teaching schedule at Caltech—but he hardly slowed the pace he had adopted as a young man. On his birthday in 1958, he had written in one of his diaries, "Life starts at 60." Did he really believe that? For his generation, which weathered war and privation, sixty was an advanced age. But the fire of ambition, pride, and his belief in his special calling did not let up.

Now, as he approached mandatory retirement age at Caltech, he was still in demand as a public scientist and imaginative thinker. He was invited to speak and consult around the world. One speaking engagement thrilled him nearly as much as receiving the Medal of Freedom. He was invited to address the 400th anniversary celebration of the birth of Galileo, the undisputed father of astronomy. "They want me as one of a very few-selected speakers" at the 1964 event, in Italy, he wrote his daughters in the Swiss Alps.[7] He planned to speak about supernovae.

One opportunity, however, he passed on. Surprisingly, perhaps, because it was the best chance he had had in many years to renew his attack on the expanding universe. Beginning in 1958, *Science* magazine approached him with an offer to write a piece on the redshift. The magazine wanted to publish a collection of articles laying out all the evidence for and against the expansion of the universe that had been gathered in the three decades since Hubble's discoveries. Zwicky was an obvious choice, since he had been among the most prominent naysayers. Negotiations between him and the editors dragged on for two years. "Contrary to what the cosmologists maintain," he wrote,[8] "all the models of expanding universes so far proposed are increasingly challenged by our research."

The "our" referred to Milton Humason, Hubble's old collaborator and by this time a friend and confidante of Zwicky's. His idea, similar to what he had argued decades before, was that the universe was static and in thermodynamic equilibrium. Zwicky also invoked his old tired light theory that "gravitational friction" might cause the redshift. But he never wrote the article, and *Science* never published it. Was it possible that Zwicky had privately come around to the point of view that had gradually taken hold among most other scientists, that the universe was indeed "blowing up?" But that he simply could not admit it?

As he told his Swiss friend Paul Wild, he knew there would be a time when someone else was right. He never did publicly admit his error, but

over the years, he began using redshift data in his own velocity calculations, though always qualifying it, by using the word "apparent," the same
one Hubble himself used.

In September 1963, while on business in France, Zwicky met Yuri
Gagarin, the first man to fly in space. Gagarin beat the American astro-
naut, John Glenn, by almost a year, circling the Earth once in his Vostok
1 capsule. His courageous 89-minute flight made him the first to see the
planet in all its blue finery.[9] He was less suited to the international fame
that came his way after he landed. He seemed uncomfortable, even dumbstruck, in front of the cameras. But he opened up to Zwicky about Soviet
intentions in space. It didn't hurt that Zwicky had learned to speak Gagarin's native language.

"He is a small man, very pleasant, kind of peasant-like,"[10] wrote Zwicky.
Only two weeks before the two men met, President Kennedy announced
that the United States would put a man on the moon before the end of
the decade. He said he would welcome Soviet participation in a joint program to meet the target.

"In spite of President Kennedy trying to make everybody believe that
the Americans and the Russians will go to the moon together, nobody in
Paris believed that," Zwicky said. "In fact, Gagarin stated that the Russians will go alone. I'm sure he speaks for the Russian government."

At the time, it must have seemed to the young cosmonaut, as it did to
many space enthusiasts around the world, that joining up with the slow-
footed Americans would just hold the Soviets back. If that was his
thinking, he was soon proved wrong. Gagarin's flight turned out to be a
high point for the Soviet space program, while the Americans rushed
ahead with the Apollo launches that landed Neil Armstrong and Edwin
"Buzz" Aldrin on the moon in 1969. Gagarin had died a year earlier, in a
plane crash.

Two months after his trip to Paris, Zwicky returned from a weeklong
observing marathon at Palomar with what he modestly termed "some startling discoveries. . . . Most of the astronomers," he wrote gleefully to
Fran, "will not be happy about my always picking up new things."[11]

There was a lot of truth in his boast. At this time, he no longer concerned himself with the big, overarching problems of the age and fate of
the cosmos as a whole, the topic that had been engaging astronomers and

physicists since Hubble's discoveries. Though he had once theorized on changing constants and dark matter, by late middle-age, Zwicky had become more of a phenomenologist, digging around in the rubble of the universe to unearth curious objects. Like a dinosaur hunter with a knack for digging in the right places, Zwicky had an eye for the strange and out-of-place.

"Fritz was very good at finding weird things," is the way Princeton's Jim Gunn put it.[12]

The most significant of these late-in-life researches concerned a new class of galaxies, which he called compact. Compact galaxies were small, dense objects that seemed to differ greatly, and not just in size, from big galaxies like the Milky Way, with its sweeping spiral arms 100,000 light-years across. Zwicky said the compact models could be as small as a hundred light-years in diameter.[13] He had originally become interested in compact galaxies "in the hope that they would act as gravitational lenses and thus image around them some exceedingly distant galaxies." While he never succeeded, the subject intrigued him enough that he began a systematic search for compact star systems. In 1964, he published a list of compact galaxies, which received a lot of attention.

"You can rest assured that the catalogue contains fundamental data on which very many people will depend for a very long time," wrote the astronomer, D. W. N. Stibbs, of St. Andrews University, Scotland.[14]

Caltech's George Djorgovski agreed that Zwicky's interest in compact galaxies was an important insight, aiding the understanding of how stars are formed, and where. "You can have some (galaxies) very diffused, and to the other extreme very compact. Most of those Zwicky picked up were star-forming ones." Zwicky speculated on how they might develop and suggested later on that they might contain black holes, which he called "Objects Hades."[15]

These late-in-life discoveries solidified Zwicky's new reputation as an expert on galaxies, particularly clusters of galaxies. In 1961, he began work on his magnum opus, a six-volume catalogue of galaxies and galaxy clusters that would take almost a decade to complete. A list of more than 40,000 objects, the *Catalogue of Galaxies and of Clusters of Galaxies,* known as the Blue Book, consumed the combined efforts of all his assistants as well as his family. It was an effort worthy of a life's work, and its value is shown by the fact that it is still being consulted.[16]

One of his compact galaxies, known in the literature as I Zwicky 18, has been a subject of study and debate ever since he found it. Some 59 million light-years away, it is almost devoid of elements heavier than hydrogen and helium, which astronomers think makes it a snapshot of the early universe, before stars began kicking out the heavier elements through supernovae and other mechanisms. The relative abundance of helium is quite high, however, which suggests the gas was made in the Big Bang itself. "This is a very important line of reasoning in support of the hot Big Bang," according to Harvard's Robert Kirshner.

Zwicky was also much puzzled at this time by another set of strange objects in the sky, which he called faint blue stars. He and his friend, Milton Humason, had produced a list of forty-eight such stars, which came to be known as Humason-Zwicky stars, as early as 1941. The fact that they were blue meant they were very hot. In the sixties, he did a lot of work on the subject with William Jacob Luyten, a Dutch astronomer at the University of Minnesota, and the Mexican astronomer Guillermo Haro.

While Zwicky worked at Palomar, Haro compiled an exhaustive list with Luyten of 8,746 blue stars in the northern sky. Haro used the 28-inch Schmidt telescope at Tonantzintla, 7,000 feet up a mountain in the Mexican state of Puebla. It was a sibling of the big 48-inch Schmidt telescope that Zwicky was now using at Palomar, when he couldn't get time on the 200-inch Hale. In 1959, Luyten proposed the term "dwarf nova" as a "cover-all for the time being."[17] Although neither Zwicky nor his collaborators suspected it, they were on to what would become one of the most important astronomical discoveries of the twentieth century.

All three men shared the same interest in sniffing out the strange and unusual in the heavens. By 1964, they had helped raise the profile of these strange blue stars so much that an international conference was held in Strasbourg, in the Alsace region of France, to share findings and thoughts on their origin and composition. Zwicky was of the opinion that they were located in our galaxy. For a few dozen very important ones, that would prove to be a critical error.

Zwicky's work with Haro and Luyten exemplified his willingness, even eagerness, to seek out colleagues who were not part of the established hierarchy in astrophysics. Later on, Haro, partly on the basis of the work he did with Zwicky on blue stars, would become the first astronomer working in a developing country to be elected to the Royal Astronomical Society in Great Britain.

Zwicky and Luyten became close. Whenever the Dutch astronomer visited California, he made sure to drop in on the Zwickys in Pasadena. From their voluminous correspondence, it was clear that Luyten looked up to Zwicky as a model of the fearless, unfettered investigator. "He was the closest approach to a universal genius I have ever met," Luyten wrote Margrit years later. He recalled the "many occasions in your house in Pasadena," where they discussed everything from the physical sciences to scientific politics.[18]

Luyten especially looked up to Zwicky for solace and advice. He needed plenty of both because he, like Zwicky, felt he had gotten a raw deal. In particular, he believed that Jesse Greenstein, head of Caltech's Astronomy Department and Zwicky's superior, had stolen his thunder on white dwarfs.

Unlike a supernova, a white dwarf is the remnant of a less massive type of star, ranging from the mass of our sun up to about eight times more. When such a star exhausts most of its nuclear fuel, it shrinks to a very hot, very small, core, though its radius is still a hundred times greater than a neutron star. It is now known that a white dwarf can produce a supernova, but it needs to feed off a companion star to do so.

Luyten had done a lot of work on these shrunken objects in the fifties and early sixties, yet felt he had been elbowed aside by Greenstein. Of course, Greenstein had at his disposal the big telescopes in California, as well as the reputation of Caltech as the premier astronomy teaching and research center in the world. "Did you see the recent *Sat. Eve Post* article," Luyten wrote Zwicky in October 1959. "He has now appointed himself the sole authority on White Dwarfs. This is the second time that he writes a long article on White Dwarfs and manages not to mention my name once. . . . After I find a star . . . determine its color . . . to be white . . . get the magnitude and send him an observing chart . . . then finally he 'discovers it.'

"The worst, however, is his calling himself a stellar undertaker. Baade called me that many years ago. . . . But now apparently Greenstein is everything in White Dwarfs. Well, in future he can go and discover his own—without observing charts from me."[19] Greenstein the Great was how Luyten began to refer to him.

"Concerning your troubles with all the stinkers who deliberately usurp all of your discoveries and achievements, I heartily sympathize with you," Zwicky replied. "As a lone wolf, the only chance you have is to write some books."[20]

That is what Zwicky was doing. In a few years, he would publish one of the most notorious books in recent scientific history, settling all his scores. But Luyten didn't have Zwicky's unbounded self-confidence, nor did he take as much pleasure in a good fight.[21]

Sorting out the truth of scientific feuds many years later is difficult. Greenstein was a respected scientist who did important work on his own. In his position as head of astronomy at Caltech, he was a capable, discerning administrator. He was, of course, like all scientists trying to write their names in history, competitive. But after Zwicky passed away, while others seemed to do all they could to extinguish his reputation, Greenstein offered one of the most perceptive assessments of Zwicky's work. (Of course, that was in a public setting, at a memorial event. In private, he admitted, "I disliked him as a human being." "He was vain and very self-centered."[22]) Like every administrator who supervised the Swiss, Greenstein had felt the sting of his sharp tongue.

Luyten did finally get the recognition he was looking for in 1964, when he was awarded the Watson Medal, which honors important contributions to astronomy. The award cited Luyten's work on white dwarfs. He suspected, correctly, that it was only through Zwicky's intercession that he was given the award. He wrote a grateful letter to his friend.

Zwicky replied that the fact that he was able to influence the outcome, despite being "heartily hated" in many quarters, showed that "they are not all hardened criminals in the National Academy of Sciences."[23] In his letter to Margrit, a decade later, Luyten claimed that all "the recognition I have ever had came through Fritz."

While all this was going on, while Zwicky was consumed with his galaxies catalogue and puzzling over his strange blue stars, a colleague was about to show the world the greatest object of cosmic disaster since the supernova. The fact that the discovery was one which Fritz came close to making himself, only to see it snatched away by a man who worked nearby on the second floor of Caltech's Robinson building, only made it more ironic.

Maarten Schmidt was born in Groningen, the Netherlands, in 1929, three decades after Zwicky, making him a member of the new generation of astronomers and astrophysicists who were building on the work of Fritz's pioneers. He was not yet a teenager when German Messerschmitts appeared in the skies, beginning an occupation lasting five years and subjecting the population to fear and starvation.

The single benefit to the blackouts enforced by the Nazis was the black-as-pitch night sky, revealing the heavens in their undisturbed glory. While Baade, at Mount Wilson, took advantage of similar blackouts in California to double the size of the universe, Schmidt's evening walks with his father, an accountant, enabled him to witness the cosmic panorama even in the heart of the city.

The boy's interest took what he called a "quantum leap" when he peered through the small telescope his uncle kept on the upper floor of his pharmacy. This inspired the 12-year-old boy to make his own, using a lens he scrounged from his grandfather's home north of Amsterdam.[24]

Once the Nazis were vanquished, Schmidt went to the local university, where he studied physics, mathematics, and astronomy. His talents came to the attention of the nation's leading astronomer, Jan Oort, an acquaintance of Zwicky's, who would give his name to the Oort Cloud, a teeming region of objects in the far wilderness of the solar system. Oort believed, correctly, that many of the comets that appear in the heavens, including Halley's famous object, originate in that far-off region.

Schmidt's early work involved researching the spectroscopic emission lines of hydrogen, learning its telltale signature like a forensics expert examining a crime scene. By studying the genomes of light from various places in the Milky Way, Schmidt was able to chart the spiral arms of the galaxy. But it was his familiarity with the signature of hydrogen that would prove crucial a decade later, when he discovered a type of object that made a supernova look like a child's cap gun.

By this time, another important instrument had taken its place in the astronomer's toolbox: the radio telescope. For millennia, visible light was the only medium humans used to understand what was going on outside Earth. But as far back as the middle of the nineteenth century, James Clark Maxwell showed that electromagnetic waves come in all sizes. They range from the shortest wavelengths, and the highest frequencies—powerful gamma rays and X-rays—through ultraviolet, visible, infrared, micro-waves and, finally, low-frequency radio waves. Radio waves are very

Maarten Schmidt's groundbreaking work on quasars earned him lasting fame. When he described his discovery to Zwicky, his neighbor in Caltech's Robinson building, Fritz thumped himself in the head, realizing he had been close to making the discovery himself.

much longer than light waves, ranging from centimeters to miles, which is why radio telescopes have to be so large. You need a big dish, or a series of dishes spread out over a wide area, to accurately capture and catalogue the emanations from space in the radio frequency range. That can be a problem, but a radio telescope has several important advantages over an optical telescope: it can be used in daytime; it is not affected by atmospheric distortion; and because radio waves are so large, it can see through interstellar dust that would block radiation with shorter wavelengths. That meant radio telescopes could plumb extremely distant regions of the universe. By the sixties, radio telescopy had come of age. Instruments in North America, England, the Netherlands, and Australia were identifying hundreds of bright radio sources.

Meanwhile, Maarten Schmidt's reputation had grown so substantially that in 1959, Jesse Greenstein offered the 30-year-old Dutchman an associate professorship at Caltech. In 1961, he got his chance to use the big Hale telescope on Palomar Mountain, where he began probing the mysterious radio sources. But it wasn't as easy as pointing the telescope at the sky and nailing it. In the eyepiece of the 200-inch telescope, the sky is a very congested place. And because radio waves are a million times larger than light waves, they produce blurry images. Tracking a radio object to its visible light source was a laborious, time-consuming process, but Schmidt kept at it.

Most of the radio sources he looked at turned out to be ordinary elliptical galaxies. Zwicky presented the results of one curious object at the Strasbourg conference on Faint Blue Stars. HZ (Humason-Zwicky) 46—the forty-sixth of forty-eight faint blue stars the pair had investigated in 1941—he said was "a supergiant galaxy with a very compact core."

But a few of the objects were puzzling. They didn't look at all like galaxies. Instead, they looked very much like stars. Very powerful, very blue stars. These objects were very much on the mind of Maarten Schmidt.

Like others, Maarten Schmidt had a rocky relationship with Zwicky.[25] Once Zwicky objected to a visitor using a piece of equipment and, when Schmidt stood up for the interloper, Zwicky lost his temper and kept his door shut for a while. Schmidt is tall, about 6 feet 2 inches, and remembered Zwicky as being about the same height. He recalled two things in particular about his colleague and office neighbor: He had a lumbering way of walking, "like a man from the mountains," which of course he was;[26] and the frequency with which he used the word bastard. By this time, Zwicky had begun using a particularly colorful term for those he disliked, "spherical bastard." Schmidt denied the story told by another scientist that Zwicky threatened to kill him.

One of the radio sources Schmidt was particularly interested in was named 3C 273.[27] The 210-foot diameter Parkes radio telescope in Australia was able to narrow down the field enough that Schmidt felt he had a chance to spot it with the Hale telescope one night in late December 1962. This was just weeks after the end of the Cuban missile crisis; while the world breathed a sigh of relief at avoiding nuclear disaster, Schmidt buckled down on a California mountain top.

"It was romantic!" he told an interviewer later. "Once in a while you just had to stop and look around you. At times it could be damned cold, but I had a hot suit"—a pair of electrically heated coveralls apparently left over from World War II. "It said 'Army Air Forces' on the front, and it did a rather good job of keeping you warm."

As it turned out, the mysterious 3C 273 was really two sources, a thirteenth magnitude star and an attached jet of gaseous material. The brightest stars in the sky, such as Sirius, Antares, and Vega, are first magnitude stars. The faintest stars visible to the naked eye are sixth magnitude, so 3C 273 was much dimmer.[28] But it still outshone the other radio galaxies Schmidt was used to working on. When he finally got good images on photographic plates, the spectra made no sense. There were some broad emission lines on the spectrogram that, once again, didn't match anything he knew. Or anyone else knew, for that matter.

On a Monday, a few weeks later, Schmidt was sitting in his office, adjacent to Zwicky's, and looked again at the image in his viewer, when something clicked. The image, he suddenly realized, looked a lot like the fingerprint of hydrogen, the primary fuel of stars. Only it was redshifted tremendously, which meant it must be traveling away from Earth fantastically fast, almost 30,000 miles per second, and was fantastically far away, about 3 billion light-years. Yet it was brighter than most known galaxies much closer.

If it was that far away, how could it even be seen? It shined with the light of 2 trillion stars, hundreds of times brighter than our own galaxy.

And yet, it could not be a galaxy. It was only about the size of our solar system, less than a light-year across. What was going on? In his excitement and confusion, Schmidt went looking for someone to talk to about it. He found Jesse Greenstein, who had been puzzling over a similar object, an even fainter, sixteenth magnitude object, known as 3C 48.

In his oral history, Schmidt said that when he looked at Greenstein's "star," he saw a redshift even greater than Schmidt's object. "They mutually confirmed each other," Schmidt said.

The two men spent the rest of the afternoon trying out alternative explanations that would show these objects were much closer, and in our own galaxy. Everything they knew about the makeup of stars said they had to be. Perhaps they were made of some unknown element. After the workday ended, with no good explanation, Schmidt went home and said to his wife, "Something terrible happened at the office."[29]

It turned out that it wasn't terrible at all. Maarten Schmidt had discovered the quasar (quasi-stellar radio source), an engine of almost unbelievable power. So unbelievable, in fact, that it took astrophysicists another six years of puzzling before Donald Lynden-Bell, a former postdoc of Schmidt's, helped come up with what is now the accepted explanation: a hungry black hole in the process of consuming a meal. Although nothing can escape the fearsome gravitational power of a black hole, material at the edge of the black hole's whirlpool—called an accretion disc—is heated so much that bursts of energy are shot out at close to the speed of light.[30]

This energy blast is what the radio telescopes on Earth were picking up and what Schmidt saw when he finally nailed down 3C 273. The reason it appeared as a radio signal is that, in the presence of the strong magnetic field that surrounds a black hole, electrons being spun off at high speed move along paths that resemble a helix—like a stretched-out slinky—giving off radio waves as they move. It wasn't a galaxy, and it wasn't a star or even a black hole, exactly. It was the radiation spun off by the most powerful dinner show in the universe.

Schmidt recalled telling Zwicky about his discovery. "It was very interesting. When I discovered quasars, I talked to Fritz. I said there were these blue objects at high (northern) latitudes. They are not stars. He stood there and realized he had been very close to discovering them himself. He hit his head with his hand."

Schmidt's discovery made him famous. His youthful, angular, bespectacled, face graced the cover of *Time* magazine.[31] It was all very flattering and, not insignificantly, quite good for his career. He became chair of the Division of Physics, Mathematics and Astronomy at Caltech in 1975, and then director of the Hale Observatories.

"It was a fantastic event," Schmidt said. "But once it's done, it's done."

The more satisfying work came later, when he was able to show where quasars fit on the family tree of the universe. After years of painstaking work, he was able to determine the population of quasars "as a function of cosmic time." As some of the most distant objects that can be studied, which also makes them the oldest, "they show a snapshot of what the universe was like at that time," he said. "I was able to collect evidence about the early evolution of the universe."

Quasars are cosmic dinosaurs, ancient beasts roaming the landscape of space and preying on weaker creatures to feed their enormous appetites. This fact proved to be an important nail in the coffin for the so-called steady state theory of the universe, which maintained that the universe had always been like it is today and always would be.[32] That would mean the universe should be the same all over. These relics of an ancient cosmos, so fantastically far off and so different from anything being created in space today, were proof that the young universe was a far different place.[33] The universe was indeed expanding, growing older, and at breakneck speed. No matter what Fritz Zwicky believed.[34]

It is now believed that there are supermassive black holes at the center of most large galaxies, like the Milky Way. But these days relatively few have quasars, or what are now called active galactic nuclei. They are active because they are eating. Over time, the vast majority of black holes consume all the dust and gas and other things in their region and go into hibernation. The black hole at the center of the Milky Way, Sagittarius A*, is one of these hibernating black holes. There is a distinct possibility, however, that this cosmic beast could start eating again one day. The closest big galaxy, Andromeda, is steadily approaching the exurbs of the Milky Way. The two giants will butt heads in about 4 billion years. That event will send tides of gas and dust in space washing up against the lethal shoreline of the black holes in both galaxies, awakening their hunger once again.

It should be quite a show for any observers who happen to be around. They won't be on Earth. In a few billion years, the sun will exhaust its own fuel supply of hydrogen and begin bloating, and expanding, and turning redder. As it grows, it will boil off Earth's oceans and destroy life. Perhaps by then, scientists will have figured out a way to follow Zwicky's advice and create new, habitable planets.

16

BANISHMENT:
EVEN THE OLD LION MUST ROAR

SCHMIDT'S DISCOVERY OF THE quasar set the stage for one of the most pyrotechnic feuds of the era, one remarkable even by the standards of Zwicky's many eruptions against the grey thinkers and spherical bastards. Fittingly, his opponent was a man of similar stature in science and one whose ego could expand to fill the available space every bit as successfully as the Swiss astrophysicist.

Allan Sandage had been one of the first astronomy students at Caltech after the department was created. Like Maarten Schmidt, he was one of the new generation of astronomers, almost three decades younger than Zwicky. Though he liked to describe himself as a hick from Iowa, Sandage had a strong personality of his own and stronger opinions. "You weren't anybody in astronomy if Sandage hadn't stopped talking to you at one time or another," one science writer said of the man considered to be one of the giants of astronomy over the second half of the twentieth century.

After Hubble's death in 1953, it fell to Sandage to try to narrow the Hubble Constant for the age of the universe. Baade's measurements in 1952 put it at 6 billion years. In 1965, Sandage, using better methods, came up with a range of between 15 and 25 billion years. Only the Hubble Space Telescope would do better.

Like Schmidt, in the early sixties, Sandage was fascinated by radio sources. Soon after Schmidt's discovery of quasars, it was found that their optical counterparts were extremely blue. This led some astrophysicists to conclude that perhaps quasars could be identified by their color alone.[1] Sandage began to study these blue stellar objects and soon realized that many of them were not associated with known radio sources. He believed he had found an entirely new population in the heavens, which he called "quasi-stellar galaxies."

In May 1965, he submitted a paper to the field's premier publication, the *Astrophysical Journal,* which was immodestly headlined "The Existence of a Major New Constituent of the Universe: The Quasi-Stellar Galaxies."

In his paper, Sandage said he had found evidence that "most of the (faint) blue, starlike objects" being found in the northern sky were extra-galactic [outside the Milky Way] and represent an entirely new class of objects." Further, he said, the objects "would seem to be of major importance to the solution of the cosmological problem," meaning the shape and age of the universe. The quasi-stellar galaxies, he said, were far denser than the quasars, which meant they had not yet evolved into conventional galaxies. These objects, he said, revealed the way the universe looked 90 percent of the way back to the "creation event."[2]

If Schmidt's discovery of quasars had uncovered the age of the cosmic dinosaur, Sandage seemed to have found their aquatic ancestors.

Allan Sandage, who helped narrow the range of values for the Hubble Constant for the age of the universe, engaged in a memorable feud with Fritz Zwicky, who accused Sandage of stealing his discoveries. Despite warning that their dispute was harming the reputation of the Palomar and Wilson observatories, their superiors were unable to stop the scientific bloodletting.

Subrahmanyan Chandrasekhar, the future Nobel Prize winner who contributed to the understanding of the life cycles of stars, was the managing editor of the *Astrophysical Journal* at the time. He thought the paper was so important that he didn't submit it to any referees for review. In an extraordinary step, he also delayed publishing the May issue in order to get the article in.

Sandage's was a sensational claim, and it immediately came under attack, principally from Fritz Zwicky, who felt that Sandage had simply stolen his work on compact galaxies and faint blue stars, and given it a spectacular twist.

"All of the five quasi-stellar galaxies described individually by Sandage (1965) evidently belong to the subclass of compact galaxies with pure emission spectra previously discovered and described by the present writer," he wrote.[3]

His private correspondence reveals how wounded he was. In a letter to his friend William J. Luyten, Zwicky said Sandage was "trumpeting around that he had made the biggest discovery in a century, which turned out to be nothing else but what I had said all along." He also said Sandage had been bragging about holding up publication of the *Astrophysical Journal* so his "big discovery would be immediately announced and he also seems to have mobilized all of the news services to make his earth-shaking announcements."[4]

Zwicky wrote that after Sandage's article appeared, he intended to send an "immediate reply which no doubt will be difficult and start again one of the most colossal brawls."[5] There was no hint of regret about it. He relished the combat.

Zwicky was not only angry at Sandage; he never forgave Chandrasekhar, pointing out that while the editor rushed Sandage's paper into print, he had rejected Zwicky's own article about compact galaxies the year before. "In sharp contrast to their ready and uncritical acceptance of all sorts of childish phantasies [sic] and stolen ideas," he wrote later, "the editors of the *Astrophysical Journal* exhibited an almost unbelievable lack of tolerance and good judgment by rejecting my first comprehensive and observationally well documented article on compact galaxies."[6]

Zwicky's work had been unveiled at the April 1963 meeting of the American Astronomical Society,[7] where he argued that the radio stars were at the "most luminous end of the sequence of compact galaxies." Yet his follow-up paper had been rejected by Chandrasekhar. Zwicky saved the rejection letter of April 2, 1964, to the end of his life. "Communications of this character are outside the scope of this journal," it said.

Sky & Telescope, the leading popular magazine on astronomy, also carried a piece on Sandage's discovery. In his subsequent letter to the editors, Zwicky began by complimenting them for noting correctly that he had previously reported on the existence of "very compact galaxies, some of them blue, and that I anticipated their relation to quasars." But he said they failed miserably by allowing Sandage to claim to discover the quasi-stellar galaxies. He said he had in fact "stressed on many occasions that many of the" blue objects he and his associates had found were "compact, blue extra-galactic galaxies, most of them radio quiescent."

Sandage's measurements of the objects' spectra "resemble in every respect . . . some of the many blue compact galaxies on which I have reported," Zwicky said. "There is thus no discovery of any 'new major constituent of the universe.'"

Princeton's James Gunn, a student at Caltech at the time, knew and respected both Sandage and Zwicky. He said it was understandable that Zwicky felt his work had been purloined. He had indeed prepared the way for both Maarten Schmidt and Allan Sandage. But he does not think Sandage was guilty of scientific theft. "It is true that Fritz knew about these [quasi-stellar objects] but it is not true that Sandage hadn't found something new," Gunn said.[8]

According to Gunn, Zwicky did not realize how far away his objects were, or what they truly were. "He did not know about the redshifts." And when it came to these strange, quasi-stellar, radio-quiet objects, the redshifts were everything.

Naturally, Sandage took his discovery to the physics colloquia, which took place once a week during the school year. These meetings, at a time when Caltech was leading the world in astrophysics, were heady gatherings that featured some of the greatest scientists presenting their latest evidence and theories. Sandage surely must have known that Zwicky would be attending. As a respected elder statesman, Zwicky was accorded the privilege of sitting in the place of honor, in the front row. When Sandage appeared, the auditorium was filled with 200 scientists and other interested parties. It's unclear how many expected the fireworks that ensued, but it was quite a show for those on hand.

"Zwicky was very upset," recalled Gunn. But he waited. "Fritz let him go all the way through his presentation before he stood up."

"What's new?" Zwicky shouted, according to another observer. "I discovered all this years ago." He lashed out at Sandage, saying the so-called "major new constituent of the universe" was simply the blue stars and compact galaxies he and Humason had found earlier, and which were investigated in detail by Luyten and Haro.

According to witnesses, Sandage seemed to shrink before Zwicky's assault. Though Sandage was a huge figure in cosmology, he was over-awed by Zwicky, according to the physicist Richard Ellis.[9] Even if Zwicky didn't feel Sandage had stolen his work, he likely wouldn't have liked him, "because he was so conventional," Ellis said.[10] The fact that he had, in Zwicky's mind, committed scientific theft, caused Zwicky to unleash his full arsenal of anger and contempt.

Maarten Schmidt recalled an amusing incident with Zwicky at a different colloquium, this one involving Richard Feynman, the legendary rascal who scribbled equations on placemats at Los Angeles strip clubs and once peed while standing on his head to prove that urination is not caused by gravity. Feynman began his teaching career at Cornell, but after the war was recruited to Caltech, where he became acquainted with Zwicky's outbursts. Schmidt recalled Feynman pausing before making his presentation, after spotting Zwicky in his usual seat.

"Feynman was very clever," Schmidt recalled. "'Before I start,' [Feynman] said, 'everything I'm going to say is known to professor Zwicky.'" Schmidt laughed. "Fritz looked very happy."[11]

"It was meant sarcastically, I'm sure," said Virginia Trimble,[12] professor of Physics & Astronomy at the University of California, Irvine. She knew both men and posed for Feynman when he took a detour into art in the sixties. Like Zwicky, who drew landscapes as a youth, Feynman was a modestly talented amateur painter. Trimble also spent a memorable evening at Zwicky's home. Beforehand, he asked whom she would like him to invite to even out the table. She suggested a brilliant postdoc at Cambridge named Martin Rees, who would go on to become a leading cosmologist. Rees accepted the invitation.

It was a pleasant evening, and Zwicky was voluble and expansive. But Trimble was well aware of the combustibility he demonstrated at colloquia. "Fritz could be charming with women and children and dumb animals," she said.

Did Zwicky know he had been baited by Feynman? Schmidt couldn't tell. But it likely wouldn't have mattered to Zwicky. The important thing was that his formidable status had been recognized and acknowledged.

Gerald Wasserburg, who joined the Caltech physics faculty in 1955, would sometimes sit next to Zwicky during these sessions. "He had this booming voice. 'Shit, I say that's shit,' he'd shout. I'd think, 'Fritz is in one of his moods.'"[13]

Wasserburg was "never afraid of him." The same could not be said for Sandage. "I know Sandage was terribly upset" over the confrontations, Ellis said.

As passionate an enemy as he could be a friend, Zwicky assailed Sandage in every forum available to him. Finally, in July 1965, Horace Babcock, then the director of the Mount Wilson and Palomar (later Hale) Observatories having replaced Ira Bowen, wrote Zwicky a letter, suggesting he tone down his criticism. It could harm the reputation of the observatories, Babcock said.

Like Bowen before him, who warned Zwicky that he would be laughed at if he continued to submit controversial papers, Babcock had no luck. Zwicky was affronted, claiming that if anyone had brought dishonor to the reputations of Mount Wilson and Palomar, it was Sandage. He now accused Sandage of using his position of influence at the *Astrophysical Journal* to get his work published immediately, "while the rest of us have to battle referees for months."[14]

The fight went on for many months, and Babcock proved powerless to stop it. In a letter to Zwicky in June 1966, he softened a little, saying he could somewhat understand Zwicky's point of view. "I think that practically all of us at one time or another has had the feeling that our prior work has been ignored and that many authors fail to give adequate credit to earlier investigators."

But he didn't give ground. In Babcock's Director's Report for the 1965–1966 year, he lavished praise on Sandage, ignoring Zwicky's work. "I do not intend to use the Annual Report as an instrument for resolving priority arguments among staff members," Babcock replied curtly to Zwicky's protests.

With the battle between two of his most honored astrophysicists still raging, Babcock sent a letter to Zwicky, informing him that the Observatory Committee had issued a new policy. This committee was responsible for allocating precious time on the Mount Wilson and Palomar telescopes, making it one of the most powerful bodies in the entire field of astronomy. The new policy held that when a Caltech staff member reached retirement age, he or she "will discontinue observations with the major instruments of the Observatories."[15]

Fritz Zwicky, predictor of neutron stars and dark matter, and the discoverer of a lot of other things, was being kicked off the mountain, in fact, two mountains. His access to the 100-inch and 200-inch telescopes would end June 30, 1966; that was the day Zwicky would retire as a full-time teacher, becoming a part-time instructor at Caltech. Zwicky saved the page from Palomar's 200-inch log, showing him as principal observer on June 14, 15, and 16, 1966. A hand-written note says, "My last run on the 200-inch."

Was the policy payback for the trouble he was causing Sandage and the black eye he was giving the observatories? There was no doubt that Sandage was more of a team player and much better connected with the big scientific institutions in America. Gunn agreed that the new policy was likely aimed at Fritz. "It was widely supposed that that policy that retired faculty could not get time on the 200-inch was invented for Fritz Zwicky," Gunn said.[16]

But he doubted Sandage had a hand in it, no matter how he felt about Zwicky. "Allan Sandage was such a powerful figure," Gunn recalled. But Jesse Greenstein, the head of Zwicky's department, would have had to go along with any plot against the Swiss. "I don't think Jesse's arm was susceptible to being twisted. I don't doubt the rule was invented to keep Fritz Zwicky away from the telescope. But I don't believe it was because of Allan Sandage."

However, Zwicky and Greenstein did not get along. That is no surprise. Fritz Zwicky took it almost as a moral commandment to oppose authority.

When he thinks back on his youth at Caltech, when it was guiding the world to a new frontier in cosmology, Gunn waxes nostalgic, especially about the remarkable, unbridled Zwicky. "Fritz was a crazy man," he said, affectionately. This is a common sentiment among those who saw both the genius and the humanity.

His colleague Wasserburg put it slightly differently. "He had all sorts of ideas, some of which were wacko and others were brilliant. He figured if there were neutrons, why couldn't you build a star out of them? The things he did were brilliant and interesting. I don't think he got along with anybody. But that's true of all creative people. He wasn't afraid."

Gunn agreed. Most people look at the world with blinders on that help them filter out unwanted noise, whether it is unusual ideas or challenges to their worldview. "I don't think (Fritz) had blinders," Gunn said. "That's how he invented neutron stars and dark matter. He didn't have hang-ups about the world. He was very open to finding new things. And he was very good at it."[17]

This ability set him apart. "Today, people work on huge physical things," the big cosmological questions, Gunn said. In essence, they are studying and mapping the forest, ignoring the trees. Zwicky "paid a hell of a lot of attention to the trees."

Because his career began before researchers even knew where the forest was, his attention to oddities was understandable. In this way, Zwicky was like the early European explorers and adventurers. The world was such a mystery that they looked for strange animals and plants from Earth's unknown quadrants. Like them, Zwicky was baiting traps on far-off islands for creatures that his contemporaries never dreamed existed.

"It's true that these days less attention is paid to those things," Gunn said. "I'm not at all convinced they won't become important again."

Kip S. Thorne, Caltech's Feynman Professor of Theoretical Physics and a 2017 Nobel Prize winner for gravity wave research, objected to the idea that Zwicky was only a phenomenologist. He was much more than that. "He had an amazing intuition and an amazing sense of how the universe works. He had this intuition that few people have."[18]

But Thorne also is among those who question Zwicky's understanding of the deeper laws of physics as they apply to cosmology. Thorne reached this conclusion after reading Zwicky's early papers on neutron stars. But he insists that he did not dislike Zwicky, saying that the latter's candor was refreshing. He said Feynman and Einstein had a similarly majestic intuition about nature, as it applied to the universe. The difference was that "they understood the deeper laws of physics."

Gunn knew that Zwicky was skeptical of Hubble's expanding universe. In fact, Zwicky's opposition, at the time, was not as outrageous as it appears today, when the issue seems settled. Gunn recalled that respected mathematicians spent a lot of effort in earlier decades trying to show that Hubble's and Humason's redshift measurements could be interpreted differently. All that was pretty much settled with the discovery of quasars.

"If I knew how strongly (Zwicky) felt I would have tried to convince him," Gunn said. Then he chuckled. "I probably wouldn't have succeeded. I never saw him change his mind, but he was not right about that."

Allan Sandage may not have had the stomach for face-to-face confrontation, but he was not the sort of person to simply give way before Zwicky's unrelenting assault. He and a colleague, Olin J. Eggen, struck back in August 1966, sending Zwicky a note informing him of a paper they had submitted to the *Astrophysical Journal*. "We feel that both sides of the question concerning the existence of pygmies should be discussed and offer the enclosed paper as an alternate viewpoint to your own."[19]

This referred to Zwicky's "discovery" of a new class of exceedingly compact stars, which he called pygmies. Any discovery of a new class of stars was going to generate great interest and controversy. But there is little doubt that Sandage took special pleasure knocking down Zwicky's latest work. Their paper looked at Zwicky's "pygmy" candidates and broke them down by spectrum and distance, bluntly concluding that Zwicky's pygmies were nothing but familiar white dwarfs.

"We believe there is now no observational evidence—photometric, kinematic or spectroscopic—for the existence of pygmy stars," they wrote.[20] In essence, Sandage was turning Zwicky's "what's new" attack back on him.

In November 1966, Zwicky responded with an assault of his own, which in turn generated another rebuttal from Sandage and Eggen. Pygmy stars failed to catch on, and Zwicky's accusation of scientific theft did not stick, either. But the long game was never Zwicky's style. Instinct, inspiration, and magmatic eruptions were what he traded in. And he wasn't through.

Fritz Zwicky was, as ever, a busy man as he approached his seventieth birthday. Showing that he never forgot a good idea, he tried to revive interest in his terrajets, saying they would be very useful in a future moon colony. Living on the moon, in his mind, was the first step toward his ultimate plan of spreading Earth's population to other planets. When there appeared to be little enthusiasm for his engines, he tried to up the ante, stoking Cold War fears by claiming the Soviets were ten years ahead of the United States in the "Mole-Power Race."[21]

How he knew this, he didn't say. But perhaps he was referring to the Soviet plan to drill the Kola Superdeep Borehole, which eventually reached a depth of 40,000 feet. At that depth, the Earth gets pretty hot—356 degrees Fahrenheit—as the Soviets found out. Zwicky never said how his machines would handle such temperatures, but he evidently thought his prescience in the space race would be enough get the Americans cracking on the race to inner space. It wasn't.

He was spending a lot of time at this point thinking about a more immediate threat than Soviet mole-tanks erupting from beneath the National Mall. Air pollution from the millions of cars choking southern California highways had become a serious threat to both the beauty—on most summer days one could only dimly see the nearby San Gabriel Mountains from Pasadena—and health of Los Angeles. The heavy, bilious pall hung over the region, nearly as dark some days, if not as dank, as London's killer fogs of the nineteenth century. Smog alerts became routine, forcing schools to cancel outdoor athletics and causing local governments to warn the elderly to stay indoors. Zwicky's letters during this time contained frequent references to the blight, expressed in his usual colorful language.

On one October day, he wrote to his oldest daughter that conditions were so bad "that I barely dared to leave the house. You . . . can be happy you are out of this mess. Yesterday was one of the worst days we ever had, the atmosphere literally stank and if you tried to wash your eyes you produced nitric acid right away."

"Here the smog has set in with a vengeance," he wrote again, later. "My eyes begin to hurt and make it difficult to search for supernovae."[22]

By this time, Zwicky had begun searching old photographic plates for supernovae that had been missed in earlier decades. Many of these late-in-life searches were conducted using plates from the Palomar Observatory Sky Survey, an ambitious project undertaken in 1949, using the 48-inch Schmidt telescope. By the time the survey was finished, in December 1958, nearly 2,000 plates had been taken.[23] When Zwicky found his seventy-eighth supernova, in 1972, he noted proudly that that was "more than one for every year of my life." He eventually racked up 123.[24]

Never one to stand aside when he saw trouble, Zwicky now mounted a one-man campaign against smog. In a column titled "A Morphologist Ponders the Smog Problem," Zwicky suggested that drivers on LA's crowded freeways, who insisted on driving alone, should be required to pay a tax for the privilege. He thought a dollar a day would be reasonable, which would amount to an hour's wage for many workers. It was meant to spur car pooling, but it quite naturally caused a furor among Angelenos. People wrote to their local newspaper that car pooling was impractical and would burn extra gas picking people up on the way to work.[25]

Zwicky also had a solution for Detroit's hydrocarbon-belching engines: he suggested interrupting the flow of exhaust to prevent waste products being released to the atmosphere. Automakers weren't much interested. "The automobile industry, instead of developing better engines and smog compressors, concerns itself largely with the styling of mammoth cars," he told a group called the WAGS,[26] which stood for Women Against Smog. Those vehicles "are about as useful as the towering and elaborate hairdos preceding the French Revolution, but which are much more wasteful and dangerous." The WAGS endorsed Zwicky's ideas wholeheartedly, saying they would not give up the fight for cleaner air until it was won. It wasn't until a decade later that the elimination of leaded gas and the widespread installation of catalytic converters began to ease the smog problem.

Retirement from Caltech was now approaching. In July 1966, following his last run on the Hale telescope, Zwicky became a part-time instructor. Two years later, he would retire completely, assuming the emeritus title, along with an office in the sub-basement of the Robinson building. It wouldn't be his idea. He wanted to keep on working, and observing, as long as his eyes and legs could hold out. But again, it would not be up to him.

That summer of 1966, the Greensteins threw a retirement party at their home for Zwicky, which Margrit attended. Despite the fact that the two men had an uneasy relationship, it was a nice gesture, the kind of thing the socially adept Greenstein was good at.

In September 1966, Fritz and Margrit flew to Germany for the Frankfurt Book Fair, where his newest book, *Jeder Ein Genie (Everyone a Genius)*, was getting its send-off. He was awed by the size and breadth of the event, housed in ten large buildings, exhibiting "the most stupendous number of new books on every conceivable subject," he wrote to his daughter, Margritli.[27] He said he never expected to be part of something so grand, but had begun to believe the new book would "probably find more readers than I had thought."

He hoped to have it published in America, even laboring for months to translate the book into English himself. "Now will start the fight with MacMillan's [*sic*] copy editor, because, in America publishers can never let well enough to be alone,"[28] he wrote. He referred to the big, international publishing house, which specialized in scientific and trade publications. This was one fight he would lose. No English version of *Jeder Eine Genie* ever appeared.

Back in Pasadena a month later, Fritz and Margrit were both on diets as the gala celebration of the 75th anniversary of Caltech's founding approached. Fritz had lost two pounds, and she had three to go to reach her goal, Margrit wrote to the girls.[29] The event would be held at the Huntington Hotel, and for this Fritz would exchange his bolo tie for a new tuxedo. Margrit planned to wear a "short green dress." This was the era of the miniskirt. Though she was well aware that her station as a famous professor's wife required caution and modesty, she was still young and would not give in to premature dowdiness.

Two years later, Zwicky was still so bitter about his banishment from the 200-inch telescope that he launched an attack on the Observatory Committee, the body charged with deciding who got access to the great telescopes on Mount Wilson and Palomar. "Hierarchical institutions have no place in any free society if they are established on a permanent basis," he wrote, saying the Observatory Committee was a case in point. "It is the only dictatorially appointed agency within the faculty of" Caltech,[30] and "it clearly violates the spirit" of codes adopted by the Association of

American Universities. He proposed that the faculty elect the members directly.

That attack failed to generate change. As a part-time, retired professor, Zwicky simply didn't have the status or the allies to carry the day. So it was a welcome respite when the University of Texas brought him in as a distinguished visiting professor. Instead of flying or taking the train, he and Margrit decided to drive, passing through the deserts of the Southwest where he had spent so much time just after the war, experimenting with rockets.

"Driving into Austin, we had a distance [*sic*] glance at the president's LBJ ranch," Fritz wrote to Fran. "Looks quite nice. We only have to wrangle an invitation, so he will give us one of his big Texas hats."[31] Lyndon Johnson's ranch was a sprawling, 1,500-acre affair that the president was so attached to that it was called the Texas White House. He spent a fifth of his time in office there, so it was indeed possible that he was at home when the Zwickys drove by. Bedeviled by the Vietnam War, the failure of which would cause him not to seek reelection in 1968, he would likely have been in no mood to hand out hats.

To Margrit, Texas was an alien culture. In terms of food, "Here, it's a lot of pork stomach and pork rinds and dry corn husks for making tamales," she wrote in distaste.[32]

While there, Zwicky got a letter from his longtime friend and colleague Milton Humason, retired and living in northern California. Humason, like all of the first generation of twentieth-century star hunters, was getting on in years and worried about his health. His blood pressure was so high that doctors advised him against traveling. He was also concerned for Mrs. Hubble, who was living alone after her husband's death. Humason expressed amusement, mixed with respect, over Zwicky's resistance to retirement. "I think it's fine that the University of Texas thinks you should not retire until you reach a hundred," he wrote.[33]

As usual, Zwicky made fans of the Texas journalists, who were eager to publicize his latest discoveries, whether they were truly discoveries or not. In April, the *San Antonio EXPRESS/NEWS* gave prominent treatment to Zwicky's pygmy stars, without noting that their existence remained very much in doubt.[34] Calling the professor, accurately, a "seeker and finder of strange and compact objects in the heavens," the author took note of Zwicky's prediction that some of these shrunken stars would likely be found closer than Alpha Centauri, the sun's nearest companion, four light-years off. This was one prediction Zwicky missed.

"In case no one has mentioned it to you," the reporter, Jerry Lochbaum, said in a follow-up note, "you are a pretty stimulating conversationalist."

Zwicky had been back in California only a few weeks when, early in 1968, *Nature* published a paper by a young Irishwoman, Jocelyn Bell, and her PhD advisor Antony Hewish, which galvanized the world of cosmology, proving that Zwicky's controversial prediction thirty-five years earlier had been correct after all. Bell, who was only 24 years old, had been following a strange signal, which she called a "bit of scruff," at the Mullard Radio Astronomy Observatory, near Cambridge, England.

During the summer of 1967 the so-called summer of love in America, when American students grew their hair long and experimented with hallucinogens, Bell was studying dozens of feet of nearly unreadable data every day. After a while, she learned to interpret the different sources of radio waves bouncing into the telescope. But the strange scruff kept turning up. Its signal was a most unusual one, consisting of a series of sharp pulses every 1.3 seconds. That was considered much too fast for radiation coming from a star. Its regularity, almost like a Morse Code signal, was baffling. And intriguing.

After ruling out interference from mechanisms on Earth, radar reflecting off the moon, orbiting satellites, and every other possibility, Bell and Hewish half-jokingly called the new source LGM-1, for "Little Green Men."

Messages from aliens were ruled out when Bell found another signal, very similar, but in a different area of the sky. Two separate groups of aliens beaming the same message at Earth was more than unlikely. Publication of their *Nature* paper, however, caused the popular press to go into a frenzy of speculation. Was this the long-sought proof of life outside Earth? Reporters peppered the new scientific celebrity with questions like, "was I taller than or not quite as tall as Princess Margaret, and how many boyfriends did I have?"[35]

What Bell had stumbled on, and what eventually earned Hewish a Nobel Prize, was a pulsar: a rapidly rotating neutron star. Its strong field forms a beam of radiation that sweeps through the heavens with each rotation of the star, like a lighthouse sweeping the sea. More than three decades after he predicted it, Fritz Zwicky's neutron star had been unmasked.

At the time, Zwicky didn't have much to say on the subject, preferring to let the proof of his stunning prediction speak for itself. In August 1969, however, he wrote to Greenstein, from his new sub-basement headquarters, that he had a flash of insight about what "pulsars really are." He never publicly revealed what he meant by the comment.

✳ 17 ✳

YOUNG AMERICA GOING TO POT

EVEN IN THE MIDST of his bitterest battles, his family was never far from Fritz Zwicky's thoughts. Providing for the girls' schooling at the expensive institution in Ftan was as much on his mind in the middle sixties as Allan Sandage's betrayal. Besides losing the income from Aerojet, he had also by this time given up contract work for the military. It was not solely because of his security problems. He found it a bother, especially with the labor of completing his galaxy catalogue and driving to and from Palomar. Feeling the financial pinch, he was always on the lookout for opportunities to give paid speeches. It was a welcome surprise, then, when three $75 checks arrived from Aerojet for patents he had filed in 1944, 1947, and 1950. It had taken that long for the US Patent Office to issue the certifications. "This will help pay for your skis and other items you need in Ftan," he wrote to the girls.[1] Considering the money Aerojet made from the inventions, the amount Zwicky received from his patents in rocket technology was less than pennies on the dollar.

As his oldest daughter neared her sixteenth birthday,[2] her father was feeling nostalgic. In March 1964, he wrote that after her confirmation later in the year, "you will be a great Margrit and not a Margritli any more."

"Let us reminisce and philosophize about various stages in your life and ours. There is, first of all, September 10 of 1948, when I saw you first, looking into the world with great wide-open and very blue eyes. Then came the days when you started babbling about various things. And

I especially remember how you tried your first steps in the Waldorf Astoria in the summer of 1949. . . . For mother and me it got quite lonely when we left you and Fran in (Ftan) in 1962."

He reminded her of the importance of the family sticking together, "then nobody and nothing can do us any real harm and we shall be happy and profit a lot to make our lives richer."

A year later, as Fran's own confirmation approached, Zwicky wrote a similar letter. At the time, he was high in the air, at the controls of the Hale telescope, "caught up here, because the big elevator does not work." The night assistant had scurried off to get help at the Monastery, where a potluck dinner was under way. "So I am thinking of you and of your confirmation in ten days."[3]

There are "many beautiful things in life . . . that you should enjoy while you can . . . there are some disagreeable and even tragic things too," for which one must prepare. "I think General MacArthur was right when he said that 'there is no security in life, but only opportunity.'" Zwicky concluded his note, perhaps with the arrival of rescuers to get him down from the telescope, with another paean, touching and vulnerable, to the importance of the family sticking together, "because I certainly need you all very much and, may be, you need me a little also."

He was especially anxious that his daughters not fall prey to the blandishments of teenage boys. "This summer I started to tell you girls what an old fogey like myself thinks about the relations between boys and girls," he wrote to Fran in the fall of 1965. "Actually, mother should tell her daughters, but being so young herself,[4] she naturally does not want to play the old fogey and I have to do it. Anyway, one of my observations is that plenty of boys will run after you . . . girls. So there is no need whatsoever that you" pursue boys. "The second advice is that to be really good friends, intelligent girls should only associate with really intelligent and decent boys who have shown that they amount to something, both scholastically and humanly."

He expanded on the subject in another letter. "As to boys possibly diverting your attention, remember two simple things. A) Boys who approach precipitously and insolently are also the quickest to drop the girls again. B) As a precaution I would never accept an invitation to go driving with one of these fellows."[5] His opinion may have been old-fashioned when it came to relations between the sexes, but he was very modern in his advice that they should have a profession to fall back on, so that they would not have to depend upon a man. "That is why you must become something to stand on your own feet," he wrote.

A month later, word reached Zwicky of new trouble in his extended family, what remained of it, in Europe. After his younger brother Rudi's death from heart disease in 1952, his sister, Leonie, and her husband, Stefan Staneff, assumed control of family interests in Bulgaria. Unable to leave after the Iron Curtain descended, they had borne a heavy burden for many years, trying to deal with the Communist government to reclaim their confiscated assets. They had at last managed to get Fridolin's house back, but then Rudi's widow, Elfriede, returned from Switzerland and filed suit for her share of the estate.

According to letters from Staneff, the Bulgarian court ordered an inspection of the house. During that inspection, a false wall in the basement was opened for the first time in twenty-two years. Staneff believed that behind that wall was where Fritz's father hid a hoard of gold from the thievery of the new Communist overlords. But when the wall was opened, all that was found were a porcelain dinner service, some Persian carpets, clothes, and stamp albums, most of it rotted away in the damp, underground vault.

Staneff now claimed that Rudi had made off with Fridolin's gold years before. "All the gold and money which your brother took out of here was your father's property," Staneff said. "Your father was an honest man, esteemed and respected. He left a splendid name. The same cannot be said of Rudi."[6]

Zwicky was having none of it. "As to any gold that Rudi is supposed to have swiped from father's estate, I think that that is pure imagination on your part. If such gold had existed, my father would have brought it to Switzerland himself, since he came long before Rudi."[7]

Zwicky himself received only 400 francs and a share of the house from his father's estate, which, he reminded Staneff, he had already turned over to his sister, Staneff's wife. By the end of his life, he said, his father was so poor that if he had lived a year longer, he would have had to support him.

Staneff, wounded, replied crisply. "I confirm once more, what I wrote is real and not my imagination, as my faculty [sic] is in good order."[8]

Fritz Zwicky turned 70 in February 1968. But Margrit was still a young woman, not even 40, and she began exploring other interests. "Friday, we got something. Guess what?" she wrote to her daughters in March. "A brand new car, [a] Mercury Cougar." It was a big step up from the Rambler she'd been driving, "sea foam green" and quite powerful. Years later, when American muscle cars of the sixties became popular with collectors, young people would frequently stop her on the street and offer to buy it. The new car promptly broke down on a trip to visit Nick and Tirzah in Santa Cruz, causing Margrit to sign up for an auto repair class at Pasadena City College.

The idea of women learning how to fix their own cars was such a novelty that the *Los Angeles Times* sent a reporter over to take pictures and do a story. In it, Margrit was identified only as Mrs. Fritz Zwicky, the "wife of a Caltech professor."[9]

She said she took the class to know what to do in an emergency, "and so as not to be taken advantage of by mechanics. The longer I am in the class, the more I realize how little I really know about a car."[10]

By the late sixties, with the smog problem as bad as ever, Zwicky had come to feel it wasn't just the atmosphere that was turning poisonous. Day by day, it seemed to him, the world at large, and particularly the United States, was becoming a less civilized, increasingly brutal place. This was an understandable reflex. If the era of the sixties was about anything, it was about challenging accepted norms, in political as well as cultural arenas. The stifling summer of 1965 saw the Watts section of Los Angeles erupt in flames and violence, the first of a series of civil disorders in America's big cities. "No-good anarchists and negroes now even carry guns and lots of decent people are beginning to crack their brains just what to do about it," Zwicky wrote to his daughters.[11]

President Johnson's decision to expand America's involvement in Vietnam spread the discontent to college campuses, where once-peaceful antiwar protests turned violent. "At the universities here the riots are increasing, both in numbers and violence," Zwicky wrote to his daughters. "Things are getting from bad to worse with student unrest [and] holdups [and] murder etc.," he wrote later. "In Santa Barbara the students burned down a Bank of America building. There go your dividends," he wrote sardonically.[12]

On another occasion, he wrote "here there have been some really sad developments in the schools at all levels, practically down to kindergarten; because even the tots think it is fun to come along with Molotov cocktails."[13]

And it wasn't just antiwar protests that had turned violent. Crime of all types, even in staid Pasadena, was soaring. "A 17-year-old shot it out with police in Altadena" (a bedroom community a short distance to the north) "the other day and got killed, was the son of a good family, too," Zwicky wrote.

Concerned about her safety, especially since her husband was still traveling a lot, Margrit now enrolled in a judo class. When the house next door was burglarized, she became even more cautious, telling her daughters that she was making sure to lock everything up tight at night. Gone were the days when people in Pasadena left their doors open. "Last week they had a riot at PHS," Margrit wrote, referring to Pasadena High School. The city was planning to hire fifty more police, just to patrol the schools.[14]

The students, when they weren't rioting, were not much to behold, so far as Zwicky was concerned. After a trip up to Nick's Point of Whales, where he toured the newly opened University of California campus at Santa Cruz, Zwicky wrote scathingly of the experience. "Even there, one finds smelly hippies and all these foul drug-smoking idiots. It seems young America is going to pot, by pot, to make a pun. It is not really funny any more."[15]

All this only confirmed him in his decision to return to Switzerland, where he began living half the year in an apartment in the village of Gümligen. But even there, there was trouble. Prices were high and rising, because foreigners were flooding in and buying land and houses, believing Switzerland was the safest place to invest in Europe. Zwicky considered them stupid, but "with the continuous strikes in England and Italy you can't really blame them."

In America, the student unrest, the crime, and outbreaks of violence in the inner cities reached a crisis point on June 5, 1968, when Senator Robert Kennedy was assassinated in Los Angeles. The former president's younger brother was campaigning for the Democratic nomination for president at the time.

Zwicky responded as an outsider, even though he had been living in the United States for four decades. "Here there was much ado about Bobby Kennedy's shooting," he wrote to Fran.[16] He was more repulsed than saddened by the national mourning that ensued. "The funeral was more elaborate than is being accorded any king and quite different from what Friedrich Schiller got or Pestalozzi or Einstein, or Nansen and other great men."

These were all admirable men, humanitarians, intellectuals, and, most critically, Europeans. He didn't consider the impact the event would have on his daughters, who had reached an age when their world was no longer circumscribed by their parents' values. Fran wrote a despairing letter about the terrible state of the world. Though he was certainly capable of similar sentiments, her father responded this time with a patriarchal "there-there."

"Thank you for your Weltschmerz [world-weary] letter," he wrote.[17] "Nobody expects you yet to do much about the ills of the world. Just keep on to your job and let us old fellows who have experience deal with the rotten problems."

Despite his very public hatred of Communism, Fritz Zwicky had more than a few fans behind the Iron Curtain, especially among the scientific class, who admired him not only for his imaginative insights but also because they were aware of his reputation as a renegade outsider by many in Western science. In late 1967, he received an urgent appeal from one of them, a Soviet astronomer named Boris Vorontsov-Velyaminov. Like Zwicky, Vorontsov was interested in those great, stampeding herds of stars, by now called galaxies, rather than nebulae. He eventually produced his own catalogue, which he called, in clear deference to Zwicky, the *Morphological Catalogue of Galaxies.*

"I got a heart attack and all my trips are canceled," Vorontsov wrote.[18] "I am told you have in USA a fine medicine against sclerosis of coronal arteries and stenocarditis. It is called prolonged nitroglycerine. . . . Is there some means to get it from you?" The drug was commonly dispensed in the United States to relieve the pain of angina, by dilating the vessels in the heart.

This was just the sort of emergency that Zwicky, with his broad range of contacts not only in the scientific but also in the political world, would move heaven and earth to answer. Even though the two men had never met, he went to work on the problem, quickly managing to obtain the medicine. With his sister Leonie's help, he smuggled it into Russia, where it was promptly confiscated by authorities.

Here, Vorontsov's wife leapt into action, appealing, as the Soviet astronomer put it, "ledge by ledge to the very medical top." Meanwhile, the respected scientist lay for months in a hospital bed, "isolated from the world." To relieve the man's boredom, Zwicky arranged to have a virtual library of lurid detective novels sent to him, with titles like, "She Died a Lady," and "Cautious Coquette." Decadent Western literature was much prized in the sanitized Eastern bloc.

At last, the authorities relented, Vorontsov got the medicine and returned home. "I cannot find the words (even in Russian!) to express to you and to your wife my gratitude for your extreem [sic] responsiveness," he wrote.[19] "I am in infinite debt to you in all respects an [sic] I wonder how can I find the way to do something for you."

No sooner was the crisis with his heart averted than he came down with liver disease, requiring seven more months of bed rest. Ironically, perhaps, the often bed-ridden Soviet scientist ultimately lived to a well-seasoned old age, dying at 90 in 1994. The robust Swiss would not manage so well.

Perhaps Vorontsov had an inkling of that when he wrote that Zwicky should take it easy. He was now approaching 71, and though his recovery from his own heart trouble had been a good one, "do not forget that (exertion) is always punished." In particular, he urged the Swiss to "reduce your efforts, especially when you are going to Mount Palomar. . . . The world prefers to hear upon the new discoveris [sic] from Zwicky for a longer period."

By this time, Zwicky had officially retired from all teaching assignments and lost his big office on the second floor of the Robinson building. His new quarters were in the sub-basement, Room 004.[20] As an esteemed emeritus professor, he was welcome to use this facility "on an indefinite basis," Greenstein assured him. The move had not been drama free. Zwicky had accumulated a lifetime of artifacts, including hundreds of old photographic plates that he and Margrit had begun searching for supernovae that had been missed earlier. The National Science Foundation was paying him "zero cents an hour" for a survey of supernovae since 1885.

He was also working on a catalogue of 2,300 compact galaxies. Maarten Schmidt recalled the uproar that ensued when a young assistant, given the task of helping the old professor box things up for his move, reached for an object on an upper shelf of Zwicky's bookcase. The old professor, snow-headed and wrinkled, shouted with the vehemence of a much younger man, "Don't touch that, that's a bomb." Given Zwicky's work in World War II, this seemed entirely possible, and the assistant backed away in a hurry.[21]

As he labored away in his underground vault, a new resident moved into Zwicky's dark domain. Robert Kirshner was a graduate student in 1970, when he was assigned a small room in the same sub-basement where Zwicky worked, surrounded by his photographic plates. Kirshner was an early riser, as was Zwicky. "You think of astronomers as being night owls, but he was there very, very early. He and I were the only ones in the building that early," Kirshner recalled.[22] To a lowly student, Zwicky could be "quite a scary guy. He would wear an eye-patch when he was looking at plates. He looked like a pirate."

Zwicky remained as angry about losing access to the big telescopes as he had been five years earlier. "Those spherical bastards threw me off the 200 goddamn inch telescope! Made up a special rule. Grrr. Them I could crush," he raged at the young man. That wasn't all he was angry about. "In 1933, I told those no-good spherical bastards that supernovae made the neutron stars. Now they find these damn pulsars and nobody gives me the credit. . . . Quasars? Quasars? Maarten Schmidt and his goddamned quasars. They are Objects Hades, by the morphological method predicted."[23]

On one occasion, Zwicky, apparently taking a shine to the young man, let Kirshner in on one of his secrets to success. "Always arrive before the Americans," he winked. Kirshner, who was born in New Jersey, didn't know what to say to that.[24]

Kirshner had a side job showing slides at the weekly colloquia. Zwicky "got up at every colloquium on cue," Kirshner remembered. Then the "tirade" would begin, about how he had known about all those things being discussed—whether it was magnetic fields of white dwarfs or galaxy dynamics—many years before. During Zwicky's battles with Sandage a decade earlier, these vituperations against the mandarins of science were understandable, even justified. "These set-piece speeches blasting the Caltech faculty were shockingly subversive, and wickedly amusing at first," Kirshner wrote.[25]

Eventually, however, the harangues "became familiar, then tedious, then a little embarrassing . . . eliciting inward (and sometimes outward) groans in the audience."

"He had a chip on his shoulder that was quite heavy," Kirshner said in a phone interview. "I didn't know what to make of him."

One day, Zwicky asked Kirshner what the first object from Earth to enter outer space was. Kirshner naturally said Sputnik. "Wrong!" Zwicky shouted. He then told the story of artificial planet zero. Kirshner nodded along, but was privately skeptical. Years later, while traveling in the Southwest, he decided to visit the National Space Hall of Fame in Alamogordo, New Mexico. Inside, he stumbled on a plaque commemorating Zwicky's feat. I'll be darned, thought Kirshner. The more he investigated Zwicky's claims, the more impressed he was.

"Because he was so eccentric and abrasive, he was given less credit than he should have gotten." That could almost be an epitaph.

As Zwicky worked away in his exile, he made one final appeal to Babcock. Since he could no longer use the big Hale and Hooker telescopes, he would like to be allowed to continue using the smaller telescopes at Palomar—his favorite, wide-angle, 18- and 48-inch Schmidts—to finish his surveys. In a short, blunt note, Babcock responded that he had taken Zwicky's latest request to the Observatory Committee. The age of 70, it seemed, was now the termination point for work with any of the observatories' instruments.

Babcock wrote: "I regret, therefore, that it will be impossible for me to accommodate your request."[26]

❋ 18 ❋

MAGNIFICENT DESOLATION

THE AFTERNOON OF July 20, 1969 was a scorcher in Pasadena, even by local standards. The green serape of spring was gone from the San Gabriel Mountains, the trees and grasses parched and brittle. On Oakdale Street, the Zwickys were hosing the roof of the house twice a day to ward off the heat. A few minutes after one o'clock in the afternoon, Fritz, still dressed in his pajamas, settled in his easy chair to watch television. It wasn't just the heat that inspired his clothing choice. He was recovering from his most serious health crisis since his heart attack, seventeen years earlier.

The trouble started with discomfort in his abdomen, which proved to be a blockage that was soon relieved. Of greater concern was what the doctors at Huntington Memorial Hospital found in the X-rays, an aneurysm in one of the main arteries from the heart.

The doctors wanted to treat the problem as soon as possible, putting in a "tube of Teflon or Dacron, or something like that," Zwicky wrote flippantly to Fran in mid-June.[1] He was deliberately playing down the seriousness of the danger. An aneurysm is a bulge in the vessel caused by a weakening of the artery wall. If it burst, it could well be fatal, though Zwicky's only comment to his daughters was that it "could have caused trouble while traveling." He was planning a trip to South America to attend meetings and deliver speeches. He had been tutoring himself in Spanish so that he could speak to the scientists in their language.

Despite his lack of concern, the girls flew home from Switzerland to be with their father for the surgery on July 2. The operation was a success, but the recovery, for a man of 71 years, was difficult.

Which was why he was still in his robe and pajamas that hot July day, two weeks later, as the Apollo 11 Lunar Module, piloted by Edwin "Buzz" Aldrin coasted in for a landing in the moon's Sea of Tranquility. It was the realization of everything Fritz Zwicky had envisioned and lobbied for since he first saw the Nazi rocket works in Peenemünde, twenty-four years earlier. Many thoughts must have been racing through his mind as he watched the Eagle land.

But, according to Margritli, who watched with him, while her mother bustled in and out of the kitchen, he had little memorable to say as America, and much of the world, stayed riveted on events 240,000 miles away. He sat there for hours, while Aldrin and Armstrong checked out the systems on board the Lem, as the reporters were calling the Lunar Excursion Module, and sent periodic reports back to Earth.

Just before 8 P.M. that night, West Coast time, Armstrong finally stepped out of the Eagle and onto the surface of the moon, declaring that it was "one small step for man, one giant leap for mankind." After the mission, Armstrong said he actually said one small step for "a" man, but the article wasn't picked up. He insisted, quite correctly, that the statement made no sense otherwise.

The big events of the day wrapped up with the longest long-distance phone call ever made from the White House. President Richard Nixon greeted Neil and Buzz like new friends he had just met on one of his beachside walks in San Clemente, California. "I just can't tell you how proud we all are of what you have done," Nixon said. "For every American, this has to be the proudest day of our lives, and for people all over the world I am sure that they, too, join with Americans in recognizing what an immense feat this is."

The president added that because of what they had done, "the heavens have become a part of man's world, and as you talk to us from the Sea of Tranquility, it inspires us to redouble our efforts to bring peace and tranquility to earth."

Zwicky always regarded such unctuous displays of chest-pounding by Americans with deep suspicion, and this occasion was no different. The only record he left of his reaction to the moon landing, and its meaning to the onrushing space age, was a snide letter he wrote to *Time* magazine the day after the Apollo 11 launch. Please spare us, he said of the magazine's coverage, the "meaningless drivel. . . . Get yourself a man with broad shoulders and a head on them for once, forged through the thoughts and acts of centuries, as your guide and not a bunch of saliva-dripping old maids."[2] Was this petty? Certainly, but all around him he saw evidence that the world he knew was passing away. What greater proof could there be than the fact that there was no mention in any of the coverage of the moon landing that it was the old Swiss professor who first penetrated space?

Maybe it had been, as his critics believed, a stunt. But his friend Gerald Wasserburg recalled Zwicky's and the country's anxiety after Sputnik. "We must launch something. Make a rocket and make it go," Zwicky shouted then. He had done that, no one else, and now people were walking around on the moon.

When he finally got back to work in his basement redoubt at Caltech, Zwicky left behind all thoughts of moon landings and concentrated on the tasks that he was sure would crown his life's work and seal his legacy: his final catalogue of galaxies and his search of old photographic plates for supernovae. With few paid assistants to help him, he turned to his family. He even managed to find enough money to pay his wife a pittance—fifty cents an hour—for her help.

"We are all working busily on the catalog," Margritli reported to her sister, in August 1969. "Poor mom has to read all the positions of all these galaxies." In another note, she complained that, the "plate mom and I are working on is coming out of our ears. There are at least 350 objects on it, which we have checked and rechecked."[3]

But it wasn't for naught. When Margrit found an exploding star—"All by myself," as she put it—she wrote exultantly, "You may congratulate me."[4]

"The first galaxy she looked at had a supernova," Zwicky wrote. "In 1936 I worked six months before I found my first one."[5]

Two weeks after the Apollo moon landing, Zwicky felt well enough to attend a special event at the Jet Propulsion Laboratory, to see the first pictures arrive from the Mariner 7 flyby of Mars. Happening in the shadow of Apollo, the event didn't get the recognition it deserved. The images from the two Mariner missions—Mariner 6 passed over the surface a few days before—revealed an unexpected world. Its surface was much more complex than the moon's, and the atmosphere, 98 percent carbon dioxide, was a surprise as well. This was the beginning of a decades-long process, which is still going on, to unravel the red planet's history and its potential for life. Earlier investigators, who thought they had seen canals crisscrossing the surface, turned out to not be so far wrong, after all. The Opportunity Mars rover would prove that water once flowed on the surface.

"Afterwards, we got a nifty roast beef dinner," Margritli wrote Fran.[6]

Later in the year, Zwicky made the promised trip to South America, for a conference of the International Academy of Astronautics, for which he was accompanied by his oldest daughter. "You can forget about the 'cheap' alligator purse, mom," Margritli wrote. "Down here, they are just as expensive as in Europe." Zwicky's Spanish language studies paid off when he was asked to give a one-hour speech to an audience in Cordoba, Argentina.

By this time, Zwicky's oldest daughter, having passed the Maturität at Ftan, had begun classes at St. Andrews University in Scotland, one of the oldest and most respected universities in Europe.[7] While visiting, Zwicky attended a conference and spoke at the conclusion to, as he put it, "knock them off their feet." There was some resistance from the membership to printing his more intemperate remarks, but Zwicky's friend and colleague, Luyten, was on the organizing committee. He insisted that, "I had been right too many times and I might be right again . . . anyway it was my business to make a fool of myself, if I wanted to do so," Zwicky wrote.[8]

Margritli spent two semesters at St. Andrews, majoring in English, which prepared her for a career as a proofreader, archivist, and translator.

Her younger sister, Fran, also passed the Maturität exam and began planning a career as an air hostess with Swissair. The idea alarmed her parents. Although aircraft hijackings had occurred as early as the 1920s, the seventies was the decade when hijackings for political purposes, as well as private gain, became an international concern. In 1970 and 1971 alone, there were a score of major hijackings, the most famous being the case of D. B. Cooper, who jumped out the back of a 727 over the state of

Washington with $200,000 in cash. Her parents urged Fran to cooperate if anyone attempted to hijack her plane. Fran wasn't nearly as worried about hijackings as she was anxious over the "prospect of spilling coffee on somebody," she wrote to her sister.

Fritz was at first skeptical of his daughter's forthcoming job for a different reason, saying air hostess wasn't a real career. He only changed his mind when he found out that Fran could get them airline tickets for a third the normal price.

By this time, Fran had also met a young man in whom she was very interested. She encountered the aspiring doctor, Hanspeter Pfenninger, while working as a physical education instructor at ETH in Zurich, the university her father graduated from all those years before. She and the other gym teachers were sent to the Swiss Federal Institute of Sport in Magglingen for seven weeks of training. Also attending the institute was a troop of doctors, under the command of a handsome captain, Pfenninger. Their meeting was inauspicious, nearly colliding while jogging on the sawdust track, called a *finnenbahn*. The female gym teachers were running counterclockwise, according to the rules, but the army physicians were running the wrong way, forcing the teachers to step out of the way.

"They had never been on a *finnenbahn*," Fran recalled. Strong-minded, like all Zwickys, she confronted the ranking officer, nine years older.

"I stopped Hanspeter and said, 'Would you mind turning around?'"

He apologized and turned his group. That evening, the group went down to the small town of Biel, where a dance was held. The army doctors had no cars, so they piled into the teachers' cars. Hanspeter, being shy, was not the sort to put himself forward with girls, so he had a drink and left. When the dance ended, however, he arranged to end up in Fran's Fiat 850. As he sat there, waiting to leave, a young soldier came up and said that was his seat.

"Find another car," Hanspeter said, obviously taken with the outspoken young woman from America, by way of the Alps.

Fran was not impressed by his boldness. "He must think he's special," she harrumphed inside. This hostility toward people who gave themselves airs was another trait she inherited from her father. When someone insisted on using the titular "von" when introducing himself, she recalled, Fritz would reply sarcastically, in kind, calling himself Fritz Zwicky von

Mollis. That was the name of the small town down the road from Glarus, where he would be buried.

As Fran started up, she asked Pfenninger to fasten his seatbelt. Seatbelts were a new thing at the time, and Hanspeter replied cavalierly that he never used one. If he thought the show of bravado would impress the girl, he was wrong again.

"I said, 'Get out of my car then.'" Fran smiled at the memory.

Despite the missteps, Hanspeter was not put off. He eventually won the girl's heart, after which he steeled himself to ask her father for her hand in marriage.

When Hanspeter arrived at the big apartment in Gümligen that the Zwickys made their home for six months of the year, Fritz hustled him into his office, where the two men conferred for what seemed like hours to Fran. She and her mother bent their ears at the door, listening for raised voices, laughter, anything. It might be the age of sexual liberation and cultural foment elsewhere, but in Fritz Zwicky's home, standards would be kept. There was no guarantee he would give his consent.

Finally, it was over; Hanspeter emerged with the hoped-for endorsement, and the couple announced their engagement on January 21, 1972. Fran learned that her father had asked just three questions. Was the young man Catholic? Fritz Zwicky was not a particularly religious person,[9] but in Switzerland, even then, interfaith marriages were frowned upon. The Pfenningers were Protestant, the young man said. Good, Fritz nodded. Next, he wanted to know if there was lunacy in the family, another very Old-World inquiry, reflecting the belief that mental problems were hereditary. No, nothing at all, Hanspeter answered. Last, Zwicky asked the young man if he would be able to support his wife in the manner her father had accustomed her to. "I hope so," he answered.

In the end, Zwicky welcomed Pfenninger into the family with great warmth, presenting him with his own bolo tie, with its turquoise thunderbird. Smaller than Fritz's, of course. He even allowed Hanspeter to carve the goose at Christmas. "I have a picture of my father watching to make sure he did it right," Fran laughed.

Before their marriage, Fran wanted to understand as much as she could about her future husband's work. She went to the hospital in Zurich where he was interning and volunteered in the emergency ward. "Boy did I see a lot of bad accidents," she wrote to her sister. "God, the junk I saw. . . . The worst accident was a young man who tried to commit suicide by standing under a landing plane. His head was bashed and right arm cut off." They pulled him through at first, but the next day his brain stopped working. The hospital received permission from his brother to harvest his organs. The kidneys came out first, then the eyes.

"The worst was when they peeled off his skin for homografts. God I've never seen anything so awful," Fran wrote. Still, she was glad she did it; now she knew "what it's like for HP. See you later old girl."[10]

The summer of 1971 found Fritz Zwicky in a most unlikely place, Soviet Russia. Despite his often-expressed antipathy toward the so-called worker's paradise, he had developed relationships with many of the leading scientists, who were familiar with his problems in America.

"It's absurd, the difficulties I had, and still have, with so many of the American (and Swiss) astronomers, and the Russians act and talk as if I were one of the greatest alive," Zwicky wrote to a family friend at the time.[11]

Once again, Margritli accompanied him on his tour, which took them to Georgia and then the capital, where they took a boat ride with other astronomers on the Moscow River. In the Crimea, they had a swim in the Black Sea, and ate and drank liked oligarchs. "They really laid out the red carpet for us, and the[y] wined and dined us royally," Zwicky wrote.

For Margritli, touring the forbidden land on her father's arm was the journey of a lifetime. The food was tasty and well prepared, and they had the freedom to go wherever they liked, she recalled. "We never felt watched," she said, though given the chill in international relations, they most certainly were.

They saw little of the poverty associated with Soviet Russia at the time, just two decades after the great war destroyed the economy and killed more than 20 million people. The Russians were anxious that their visitors feel at home. In Armenia, where Zwicky delivered another speech in Russian, he expressed admiration for the size of fruit on a tree outside the rooms where they were staying. That evening, when they returned from an outing, he found a big bag in the room, stuffed full of fruit. The quality of the Armenian brandy, Margritli recalled decades later, was "the best ever."

Zwicky was also, at this time, elected vice president of the International Academy of Astronautics, for which he served as chair of the Finance Committee. The honor was not only a symbol of the respect for him as a scientist, but also for his skills as a fund-raiser, the talent he had developed in his work for the Pestalozzi Foundation. "My shouting at everybody had been effective enough to induce even the Soviet Academy to give us a thousand dollars for our expenses in running the office at Paris," he wrote.

He was disappointed in the Americans, who were not contributing what he felt was their share. But even so, "we have enough to survive for another two years and if there is real danger of us gravitationally collapsing into the cosmic 'black hole,' or rather my red Object Hades, which has the proper color for a deficit, I can always use the sledge hammer on some of the Swiss."

The first black hole, Cygnus X-1, was discovered by the Uhuru X-ray satellite in 1971,[12] just as Fritz Zwicky completed work on his final galaxies catalogue.

This volume became known around Caltech, and in astrophysical circles, as the infamous Red Book.[13] In it, Fritz Zwicky would, as the mob says, settle all scores. All those he felt had stolen his ideas, hindered his research, and generally failed to recognize the full value of his achievements were called to account. The 27-page introduction was a road map of his work, punctuated by bursts of scientific bloodletting. As ever, he saw no reason not to name names. Hubble, Chandrasekhar, Sandage, Baade; the roster of those he arraigned constituted a veritable Who's Who of twentieth-century astronomy and astrophysics.

"As a shining example of a most deluded individual we need only quote the high pope of American astronomy, one Henry Norris Russell," he wrote.[14] He accused Russell of naiveté for believing that the fundamental physics of stars were so well known as early as 1927 that they could have been predicted even without ever seeing a star.

This was an example, for Zwicky, of the arrogance of too many scientists, who felt that they had arrived at final truths. Reaching back to one of his first publications, on the flexibility of scientific truth, he re-argued one of his bedrock beliefs, that "no statement that is made in finite terms can be absolute." Since the universe is a system always in change, no truth can hold forever. Even the so-called fundamental constants cannot remain the same.

His own, unique process of discovery, he wrote, was based on "my conviction there are more things in the sky than the most imaginative human mind can divine." Yet too many astronomers and astrophysicists went along with the herd, following the lead of the famous researchers at Wilson and Palomar as if they had been anointed on high. Continuing his march through time, he accused astronomers of the 1930s, meaning Hubble and Baade, of making claims that ultimately proved erroneous. And of making sure that outsiders who might challenge them would be denied access to the great astronomical instruments at Mount Wilson and Palomar. "I myself was allowed the use of the 100-inch telescope only in 1948, when I was fifty years of age," he complained.

"E. P. Hubble, W. Baade and the sycophants among their young assistants were thus in a position to doctor their observational data, to hide their shortcomings," and to convince the world of their erroneous discoveries. This was a most serious charge, and Zwicky did not provide evidence of the doctoring.

He didn't spare the journals, either, calling much of the research they published "useless trash." He got specific, referring to the "absurdities" that had been published about his favorite subject, after supernovae, the formation of galaxies and clusters of galaxies. He accused Hubble and Baade of denying the possibility that small, compact galaxies might exist. Striking at Baade once more, he said too many astronomers were "misled" by Baade's assertion that there was little material in the space between galaxies. Despite his discoveries of bridges, plumes, filaments, and other structures between galaxies, Zwicky's first papers on the subject, in the 1940s, were "arbitrarily and illegally censored by our observatory committee and withheld from publication in any of the regular American journals."

"The fact that, except for some outstanding exceptions like George Ellery Hale, the members of the hierarchy in American astronomy have no love for any of the lone wolves who are not fawners and apple polishers was made clear to me" on many occasions. He cited the example of the *Scientific Monthly* article in November 1940, in which Hubble gave credit to Baade for discovering two dwarf systems that had been discovered by Zwicky.

"Today's sycophants seem to be free, in American Astronomy in particular, to appropriate discoveries and inventions made by lone wolves and non-conformists," he continued.

Turning to neutron stars, he recalled that it was in 1933, in a lecture at Caltech, before the conference at Stanford, that he raised the possibility that an exploding star might collapse to a terrifically compact object made of neutrons. Afterward, for thirty years, "from 1933 to 1965 astronomers chose to ignore my theories and predictions." As late as 1959, the astrophysicist Alastair G. W. Cameron, a colleague at Caltech before heading east to Harvard, wrote that it had become settled that white dwarfs were "end points of stellar evolution. . . . Apparently only Zwicky has continued to believe that neutron stars were formed in supernova explosions."

Even in 1964, Zwicky noted, a prominent researcher said that his notion of neutron stars "has so far been accepted only with skepticism."

One can only guess at the satisfaction Zwicky took in throwing the words of his critics back in their faces.

Near the end of his jeremiad, Zwicky turned to Sandage, his most recent antagonist, charging that his so-called discovery of a "major new constituent of the universe" was one of the "most astounding feats of plagiarism."

Hardly any important astronomers and astrophysicists escaped Zwicky's wrath. And when the volume appeared, with support from the Ford Foundation, the effect on the Caltech campus was seismic.

On January 20, 1972, Zwicky's longtime aide-de-camp, Eleanor Ellison, the astronomy librarian at Caltech,[15] wrote to him in Switzerland that "you certainly are getting a lot of attention over the foreward [sic] to your catalog. It has become as popular here, with the students especially, as a 'best seller.'" She said she was "especially tickled by the comment of one of the new (first year) students—that you must be 'one of them'—against the Establishment!"

As for the faculty, she noted that one of his colleagues had "muttered that he wished you hadn't done it. . . . I say to anyone who implies that you shouldn't have done it, that when you reach 70 (he was about to turn 74, in fact), are retired and are world famous in your own right, that you can say and do anything that you want to—I also say that you were only telling the truth."

In February, Zwicky celebrated his birthday. In many ways, it was a melancholy time. Another member of his German club had died. This was Gustave Edmund von Grunebaum, the director of the Near Eastern studies center at UCLA. Gruenebaum was only 62 when cancer claimed him. He was the second member of the club to pass away.

Milton Humason, the diligent, meticulous, underappreciated researcher who helped Hubble "blow up" the universe and Zwicky to probe the blue stars that would show the way to quasars, was living in Mendocino, retired and ill. When he died, a short time later, Zwicky wrote a touching letter to his widow, expressing both his sorrow and his debt.[16] "When some of us physicists switched over from physics to astronomy in the beginning of the 1930s Milton was probably the only professional astronomer who helped us greenhorns along."

Humason had provided critical support for many of Zwicky's discoveries, including supernovae, Humason-Zwicky stars, intergalactic bridges, compact galaxies, and other things. Without his help, "I probably also would never have had the use of the 200-inch telescope."

Zwicky couldn't help taking a shot at the hated in-club in astronomy, even though Humason was a charter member, beloved as much by Hubble as by Zwicky. How Humason managed to straddle that divide was a testament to the old mule driver's sure-footedness, as well as his placid demeanor. "As with all great and independent men," Zwicky concluded, "I am afraid that the hierarchy in power has not and does not intend to put all of Milton's great achievements in the proper light, since they might look too small themselves."

Zwicky wasn't feeling well, himself. He was having persistent headaches that made it difficult to work. But he kept on, whipping himself to still greater action. In one of his many pocket diaries, he wrote, "Magic Directive of the Morphological Approach, Do not stop where others have."[17] This had been his lifelong policy, and he had no intention of changing now. He kept up his schedule in his sub-basement office at Caltech, surrounded by photographic plates and his books.

His door was open to any students who had enough gumption to venture into his domain. He may have been dismayed by the new generation, but he was popular with them. And because he felt no competition from them, he could be as expansive and generous with his time as they would wish. For the students, the opportunity to be in the presence of a man who helped bring the universe out of its infancy into its middle age was a rare one, and they felt appropriately rewarded by his stories of building the Hale telescope and fighting with Hubble and the others. Hale was gone. Hubble had been dead twenty years, Baade ten. Now that Humason was gone, Zwicky was among the last of his generation of star pioneers.

But his headaches continued. In May 1972, Margrit wrote that they had gotten so bad that he had to quit blinking altogether. Blinking was a process in which two photographic plates of the same portion of the sky are compared, one after the other, revealing any changes. It was tedious work, even for a young man, or woman, in good health. Zwicky became so exhausted that he finally went to the doctor, who found nothing wrong. He returned to work, but the next day Margrit got an alarming call from the university, saying her husband was "acting quite confused." She rushed over to the university and brought him home. Tests at Huntington Memorial Hospital revealed a mass on the brain, a subdural hematoma, which is a collection of blood between the surface of the brain and the thin sheet that covers the brain itself, known as the dura.

In a letter to her daughters, Margrit said the hematoma was on the right side, "and quite massive." On this occasion, there was no time to gather the family. Zwicky was rushed into surgery, where they drilled a hole in his skull and suctioned out the mass. The operation took four hours, from late morning until 3 P.M. By 5 P.M., he was in recovery and already lucid. By the next morning, his crotchetiness had returned. "He was already complaining they did not let him rest, always somebody poking at him,"[18] Margrit wrote, in happy relief.

Three days later, he was out of the ICU and back on his feet, using a walker to get around. "Daddy sends his love to you all," Margrit wrote, adding, "Greedie, Daddy wants to know how you are doing with Marlowe." This was a class assignment at St. Andrews.[19]

Yes, Fritz Zwicky was back. A week later, he wrote to Fran about his narrow escape from mortality. "My old friend Professor Borsook once said, that whatever happens to me is always on a Homeric scale. Well, a few weeks ago old Homer got into action again. In my sleep, or whenever I was not watching, he struck me on the head."[20]

Margrit added a postscript in keeping with the dangerous times. "Fran, if you are in an airport and shooting starts[,] duck and don't be a hero."

Soon enough, Fritz Zwicky was back at work, working on an article about the prospects of living on the moon, and seeking new targets for his wrath.

ON THE TRAIL OF ZWICKY'S GHOST

Mystery Unsolved

Could Soudan have found the long-sought dark matter particle? Was it at last giving up its life on the run and turning itself in? The physics community didn't think so. At least they weren't persuaded. To be fair, neither were the folks at Soudan. "Formally," Anthony Villano said, "we've said, if you fit all the data together in the silicon run, the statistics slightly prefer a situation with WIMPs [weakly interacting massive particles]."

The problem, said Amy Roberts, Villano's colleague, was that "the three events were a small sample. Unlike a situation where you have a lot of data points, like when you decide whether a piece of cheese is good to eat (odor, color, consistency, how long it's been refrigerated), or my brother is mad at me (facial expression, tone of voice, stalking out of the room)."

It was a delightful metaphor, one that Zwicky himself might have used, capturing in a very direct and simple way the practical elegance with which scientists think, at least those who actually care about communicating with people who aren't scientists.

Roberts was also a postdoctoral researcher, from the University of South Dakota. She loved the problem-solving end of physics. "I'm a hardware person," she said. Practical and no-nonsense, when asked her age, she seemed stumped, before recovering and answering, 33. Some scientists love to play the absent-minded genius. There was no guile in Roberts. She was so distracted by the problem she and Villano were having with the detector that her age was assigned to a lower tier of interest.

"There were several odd behaviors at the end of Friday," she said (it was a Monday). "The problem was a program we run to set up the detector to be in a state we need to take data."

CDMS (Cryogenic Dark Matter Search) II would soon be phased out. Difficulties with the detector were becoming more frequent. A new experiment, Super CDMS, SNOLAB (Sudbury Neutrino Observatory LABoratory), would be taking its place. SNOLAB would be housed in a new, even deeper, cave in Sudbury, Canada. At 7,000 feet below the surface of Earth, that detector should be beyond the reach of wandering radiation.

The new detector will also be looking for a wimpier WIMP. "We've opened our minds to a mass a hundred times lower," Roberts said. Why is that? "One of the biggest things is we haven't seen a signal yet" at the higher mass range. The WIMP Miracle Sunil Golwala and his colleagues had been hoping for hadn't shown up. The dark matter searchers are also convinced of the need to look for smaller dark matter candidates because of other recent experiments around the world, whose sensitivity has lately surpassed the experiment at Soudan.

According to Golwala, the new detector should be taking data by 2020. In the meantime, the tantalizing results from the silicon run at CDMS II continue to hold, even as it continues to be unpersuasive.

With everything shutting down in the Soudan mine, Jerry Meier was looking at life after retirement. He knew he'd had a good, lucky run. "I was hired because I could weld better than anyone else," Meier said. "That was 25 years ago. Now I'm running the place. I fell into this job and it was the best thing that ever happened to me."

Since 2014, Roberts and Villano have moved on to teaching positions at the University of Colorado, Denver. Also, in the intervening months and years, with the interest in the dark matter mystery continuing to grow, there has been an "explosion," Golwala said, in new technology and machinery built to trap the particle. With all this work going on around the world, it would seem only a matter of time before Fritz Zwicky's final mystery will at last be solved. Ironically, perhaps, Golwala himself, who has been in the vanguard of the determined but frustrating hunt, has become more agnostic, the longer the hunt has gone on. Even this presents an intriguing dilemma: science has proved dark matter exists, shaping the space around us like a cosmic potter's wheel. It is there. We seem to know that.[1] But what will it mean if we can't find it, can't smoke it out of hiding? Will that simply be a reflection of the inadequacy of our tools, or of something more? Fritz Zwicky's metaphor of the rainy rut comes to mind, and the image of experimenters digging ever deeper into it. Were he still around, might he launch a morphological analysis to unearth an entirely new approach?

Meanwhile, with the work winding down in Soudan, the people of northern Minnesota had the place to themselves again—the lakes, the hard mountains, the numbing cold of winter, thinking up new jokes to play on city people. Did you hear that Ely was advertising for people to paint the leaves in the fall? Seems the lovely old couple that's been doing it are retiring and heading south.

A SCIENTIFIC INDIVIDUAL

WHILE THE AFTERSHOCKS from Zwicky's broadside were rumbling through the scientific establishment, a telegram arrived from the Royal Astronomical Society of England. It informed him that he had been awarded the society's Gold Medal, one of the most prestigious awards in astronomy and in science as a whole, a crowning event for any star surveyor. The ceremony took place on February 11, 1972. Following a dinner of river trout and escalope of veal chasseur, Sir Fred Hoyle, the president of the society, stood and enumerated Zwicky's "distinguished contributions to astronomy and cosmology."[1]

The work of the Pasadena astronomers in the thirties proceeded along two avenues of research, Hoyle said, the expansion of the universe by Hubble and Humason, and the explorations of exploding stars by Zwicky, Baade, and Rudolph Minkowski.

While some of Zwicky's critics wanted to give a large share of the credit for the prediction of neutron stars to Baade, Hoyle was perceptive enough to realize that the groundbreaking neutron star and cosmic ray predictions bore the "same stamp of originality as many of Zwicky's later ideas."

Taking note of Zwicky's other early suggestion—dark matter—Hoyle noted, "This skeleton in the astronomer's cupboard rattles ever more loudly as the years pass by."

It still does.

All this had been done, all these achievements had been made, in spite of the fact that "for most of his working life Zwicky has had access to a telescope no larger than the 18-inch Schmidt which he himself built." Hoyle didn't dwell on the fact that Zwicky had been denied access to the great telescopes for years, or that he had been banished from them late in his working life. He didn't need to.

The Gold Medal was a great honor, the first time it had gone to a Swiss astronomer. It was also the kind of validation Zwicky had been wanting his entire working life. In the face of it, he was uncharacteristically at a loss for words. He said, modestly, that he hoped they had not made a mistake. Then he added, with rare insight, that he was sure there were others who deserved the award, but they were "more reticent than myself."

Around this time, a Swiss philanthropic organization asked if he would be willing to participate in a fundraiser offering introductions and interviews with leading Swiss thinkers and businesspeople in exchange for donations. This was another surprise, recognition from his compatriots, given his not-infrequent criticism of the Swiss. But then, Fritz Zwicky was no longer the same man who had suggested putting fighter planes in lakes. He had become a figure of legend around the world and even in his homeland, whose pitched battles were now seen, at least by those who were not involved, as evidence of a beguiling eccentricity. Whose stranger ideas were proof of an imagination so penetrating that the lesser minds of his enemies could not grasp their genius. No longer was the error Mr. Zwicky's. It was the world's, just as he knew all along.

Zwicky agreed to participate, and a young man named Alfred Stöckli, who would later co-author a Swiss biography of Zwicky, paid 2,000 francs for the privilege of spending a few hours with Zwicky in his Gümligen apartment.

Decades later, Stöckli recalled a very pleasant afternoon, although he had difficulty making the great man familiar with the thesis he was writing at the time, about organizational theory.[2] The energy crisis had arrived two years later, when Stöckli paid a second visit. This time, he found Zwicky stockpiling wood. Stöckli recalled a telling comment Zwicky made about being asked to consult on energy problems. "Finally, they see how important I am," he said, revealing, even then, after all the discoveries and awards, a lingering insecurity about his place in history.

Fran recalled another representative moment during that visit. When her father sat down with Stöckli, he had a flyswatter in hand. This was not unusual. Having lived through the Spanish Flu pandemic that killed an estimated 20 million to 50 million worldwide in 1918, Fritz Zwicky was a bit of a germaphobe and the sworn enemy of all flies. As the two men talked, Margrit and Margritli busied themselves slicing up a chocolate roll cake they made for the occasion.

"Don't move," Zwicky suddenly shouted. A fly had landed on Stöckli's lower leg. Before he could react, Zwicky whacked his young visitor with the flyswatter.

"Fritz!" Margrit scolded, dismayed at her husband's manners.

"I got it," he said churlishly.

Also at this time, he received a questionnaire circulated among retired college professors in America, sampling their post-academic lives and interests.[3] Asked for his feelings about retirement, he wrote that he "never thought of it as a break." Asked how he was spending his time, he said he was working fifty hours a week on research, on writing, and on behalf of international organizations.

"Nothing," he responded firmly, in answer to the question of what he would do differently if he had the chance to start all over again. There was no question in his mind what was his greatest loss in old age: Health.

"Of all the things you do . . . which are the things that are less interesting or enjoyable?" the form asked. "Dealing with grey thinkers, burocrats [sic] and general failures among men," he wrote.

He disputed several assertions most scientists would endorse. He disagreed, in particular, with the suggestion that scientists should share their findings with others. "One would ill serve the goals of man if one conversed with those skunks in science who are out for their own glory," he wrote.

He also quarreled with the idea that a scientist should take a critical approach to "all phenomena."

"One would never discover anything great," he sniffed. No better did he like the suggestion that a researcher should avoid becoming emotionally involved in his work. Emotion might be important, he said. For Zwicky, enthusiasm and passion were too valuable to sacrifice in the service of what he called the "voodooism of accuracy" and pretended objectivity.

In his eighth decade, his faculties were failing. He expressed amazement to his family at the dulling of his mind, which he was still enough of an observer to record, even if he couldn't stop it. As Zwicky turned 75, the world at large hardly seemed less threatening than it had been when the Soviets and Americans nearly unleashed atomic vengeance on each other.

On September 1, 1973, however, came one of those events that prove the stubborn persistence of hope that keep families, and nations, marching along. The union having been blessed by her father, Fran decided it was time to marry.

Her father gave her a choice of wedding gifts. He would send the couple on a round-the-world cruise, or he would pay for the wedding. Fran chose the second option. As much as he stinted on spending money cavalierly, Zwicky spared no expense on his daughter's wedding day, renting a chapel attached to an honest-to-God castle, in Spiez, a small town overlooking the lake of Thun, in the same region where he had met Margrit all those years before.

The Romanesque, thousand-year-old church, on a gentle slope above the lake, has a vaulting belltower with a pillared, three-aisle basilica, from which steps lead to the elevated chancel, six feet above the congregation. The walls are decorated with frescoes from the twelfth century, of Byzantine origin. A giant St. Christopher from 1300 A.D. stands above the pulpit.

As he walked Fran down the path to the chapel, her father handed her a slip of paper. It was an IOU she had written many years before, when she was just a girl and wanted to buy some treat. He'd kept it all that time, never mentioning the unpaid debt, a remarkable act of forbearance, given his frugal nature.

He emphasized the importance of family in his wedding speech. Like almost every speech he ever gave, he composed it as he went, urging his daughter to be morphological in her marriage, to approach problems with a completely open mind, and to act only when she understood the consequences and avenues her choices might take her down.

"A friend of my husband played the accordion," Fran recalled. They didn't have the money for an orchestra.

Whether it was because Fran took her father's advice to be morphological, or because she chose well, she and Hanspeter made a happy marriage, lasting thirty-nine years, producing two children, Ariella and Christian.

For Zwicky, his daughter's marriage was both a cause for rejoicing and relief. That was one responsibility that would be shouldered by someone else. In a letter to his friend Luyten, he gave what for him was a ringing endorsement of Hanspeter, saying he seemed to be "a regular guy . . . not one of the modern spinach brothers."[4]

This was his terminology for what people now call wimps, or wusses, encompassing the back-to-earth movement that sprang up in the seventies, when young people joined communes, mulched their gardens, and began meditating on the meaning of existence at the feet of Indian gurus. As a believer in the genius of the common man, Zwicky might have been expected to applaud this movement, but no.

"Those hippies must have irked him no end," Fran said. Their talk of peace and simplicity, he sniffed, was "*seifensieder*," soap suds. "He felt the hippies were talking about it but not doing it."

He was no happier when he heard that marchers in the streets of Zurich in those days of worldwide tumult had been heard to chant, "Zwicky, not Marcuse."[5] Herbert Marcuse was a leading theorist of the so-called New Left, who believed the technology and entertainment cultures were forms of social control. Zwicky, who never gave up his belief that science could solve age-old problems, did not consider it flattering that the attitude he adopted toward those in authority had been taken up as a matter of faith by the younger generation.

As he and Margrit grew older, their affection for each other ripened into the kind of comfortable companionship that sanctified the years of struggle that came before. In the Gümligen apartment, they began waking up in the middle of the night, as older people do, though while Fritz was in his mid-seventies, Margrit was just entering middle age.

"One night father said to mother, 'As long as we're here wouldn't it be nice to share some champagne?'" Fran recalled. "After that, whenever they met in the kitchen late at night they would share a baby bottle of champagne."

One sunny afternoon while she was showing a visitor from America around, Fran pulled her Smart car up to a café that served a delicious form of Alsatian pizza called a *tarte flambée* that she wanted her visitor to try. Inside the centuries-old building, a dozen customers sat chatting, emptying glasses of clean-tasting Swiss beer. The talk turned to her father's idea that to be happy one must find her inner genius. Had Fran ever found hers?

Cutting a slice of the white pie and taking a bite, she thought about this. Over her life, she'd served in the army, worked as an air hostess, and now worked as a lifeguard at a swimming pool. "I'm still finding it," she said. Her manner implied that for her, the question was not so relevant. She was an ordinary person, who had lived a simple life. The ideas her father grappled with concerned larger forces and grander people. She hadn't expected to understand him.

Still, she managed to answer one question her father never could. Near the end of his life, he became obsessed by the strange fact that snail shells spiral in only one direction. Northern and Southern Hemispheres, Europe to North America, always the same right-curling spiral. Wherever he traveled, he kept an eye out for left-spinning snails. Why nature preferred this anisotropy baffled him. He knew the natural world preferred balance. He had seen it over and over. In the heavens as on Earth. It didn't violate this principle without a good reason, even if most couldn't find it. He was particularly adept at figuring those reasons. But in a lifetime of discovery and invention, this was one puzzle Fritz Zwicky couldn't solve.

His middle daughter picked up his quest after her father's death. After years of searching, Fran finally found a snail whose shell curled in the other direction. The Far East is home to these nonconforming gastropods. She has no more idea why those snails are different than she does of why the question mattered to her father. Still, finding the wrong-way snails had given her satisfaction, as though she'd managed to add something to his legacy.

As 1974 approached, Fritz Zwicky remained as busy as ever. He was asked to chair a session of the prestigious American Association for the Advancement of Science. The event, to be held at the St. Francis Hotel in San Francisco, would focus on neutron stars and black holes, and the planners of the event could think of no one better suited to preside. The conference was scheduled for February 28, 1974, two weeks after Zwicky's seventy-sixth birthday. He agreed to go, but turned down talks at Princeton and Cambridge, citing financial concerns. He was surviving on "an unbelievably lousy pension," he griped,[6] while working forty weeks a year on nonpaying projects "as well as on national and international jobs for the good of man."

He told Luyten that he had an idea about the origin of gamma ray bursts—the most powerful radiation known. He intended "to write a short paper, which will get the boys mad again."[7] The old professor was delighted at the prospect of irritating his critics one more time. Even more, perhaps, than over the possibility that he might just solve the problem.

That paper was called "Nuclear Goblins and Cosmic Gamma Ray Bursts."[8] It invoked a process, and a colorful appellation, that he had introduced sixteen years earlier. "Some years ago, I investigated the problem of the possible existence of *cosmic* and *macroscopic bodies* consisting mainly of nuclear matter," he wrote, referring to neutron stars and the goblins. The neutron star prediction "may now with considerable probability be considered as having been confirmed through the recent discovery of pulsars."

Nuclear goblins, in contrast, remained elusive, he wrote. They existed only under pressure. Nuclear goblins were therefore expected to be found near the centers of massive stars, or degenerate stars like white dwarfs and pygmies. No, he had not given up on his pygmies, whatever Sandage might say. A goblin could be as small as a single neutron, or as large as the neutron star itself. The typical range, however, was about 10 to 33 feet. In his view, swarms of goblins would be distributed in the central regions of dense stars, like a water mist around a solid body. Individual goblins would from time to time be shot out from the swarm, and possibly break through the surface of the star into space, where it would explode.

The energy released by these explosions, of 780 KeV, was right in the range of the handful of gamma ray bursts that had been detected before then. The first gamma ray burst was detected in 1969, only five years before Zwicky's paper, and only sixteen had been found when he published the paper. They were found by accident, by America's four Vela spacecraft, popularly known as "spy" satellites,[9] which were designed to monitor compliance with the 1963 nuclear test ban treaty.

A leading science publication of the time noted there was no proof that nuclear goblins existed. But the author said Zwicky's knack for being proven correct showed the danger of failing to take his words seriously, as it "has forced much eating of (those) words."[10]

Zwicky, according to the article, "possessed two characteristics which are normally antithetical; outrageous heterodoxy and successful insight. . . . If the goblin model is to survive, it must take several hurdles yet. For example, their half-life has not yet been calculated. But even were the goblins laid to rest, Zwicky's ghost will continue to stalk."

Gamma-ray bursts are now attributed to the formation of rapidly rotating black holes, and the mergers of neutron stars,[11] but, as the article noted, Zwicky's ghost would continue to stalk.

Shortly before his seventy-sixth birthday, he felt some discomfort in his right groin. A trip to the doctor revealed a hernia, a relatively minor problem compared to the hematoma and the aneurysm, both of which were life threatening. The operation took place at Huntington Memorial at 2:45 P.M., on February 8, 1974. In a quick letter to her daughters, Margrit said the doctor assured her that everything went well, "that Daddy was a tough guy, and the doctor hopes to be in the same shape when he is his age. . . . The doctor said he will be home by the end of the week. Don't forget his birthday" on February 14.

Margrit said she would take the letter to the hospital for him to sign before mailing it. Indeed, at the bottom was written a very shaky "Papa." She noted that the camellias were in bloom. "I'll take some to Daddy tomorrow."[12]

Zwicky returned home but almost immediately began feeling unwell. He was rushed back to the hospital and straight into the ICU. Margrit waited in the hallway, while the doctors tried unsuccessfully to save his life. Fritz Zwicky died at 8:40 P.M. on February 8. The cause listed on the death certificate was heart attack, brought on by congestive heart failure. His body was cremated at Mountain View Mausoleum.

According to Margritli, the autopsy revealed a badly scarred heart.

His oldest daughter was in Switzerland when she learned of her father's death, from her aunt, Käthi, Margrit's sister, in a phone call. Margritli immediately booked a flight to California. Reaching Fran proved difficult. She was in Mexico, studying Spanish, while on leave from the airline. No one knew quite where she was.

"I wrote a birthday card for my father from the ruins in Yucatan," she recalled. Coincidentally, she decided to fly home at the same time. When she called her mother to pick her up at the bus stop in Pasadena, there was no answer, so she called a neighbor. "She said, 'Hi Franny, I'll tell Margrit.'"

The neighbor said nothing to her about her father, so she had no idea why her mother burst into tears when she met her bus at the Huntington Hotel.

"I ran to her and put my arms around her and said, 'Don't cry, I'm okay.'"

When her mother told her what happened, both women broke down. Fran's last image of her father was a very characteristic one. When he traveled, he always wore a suit, a fedora hat, with a briefcase in hand and a raincoat draped over one arm. The last time she saw him, he was going through passport control. He turned and jauntily lifted his hat.

Margrit went into mourning, but she was so busy answering the flood of letters and cards from well-wishers that she had little time to grieve. Nearly two hundred notes of condolence arrived at the Oakdale house. A close friend of Margrit's, who signed herself EU, recalled Fritz's attitude toward those who professed concern about the sad condition of the world but failed to do anything about it. He had no patience for excuse-making. "A few days before he died, I heard him say, 'If he can't do anything else he can write a letter.'"

One of the most touching notes was from Luyten.[13] "Fritz was an extraordinary person," he wrote. "He had a breadth and depth of knowledge in the physical sciences that was unsurpassed. The sessions I had with him at the dinner table on Palomar were something I'll never forget. He was really unique. And in addition he was a staunch friend—I never had a better one."

He praised his friend's "extraordinary" humanity, "in spite of his terrific learning . . . He was the closest approach to a universal genius I have ever met—and we shall not see his like again."

A month later, a memorial for Zwicky was held at Caltech. It was a modest group of about fifty, including Fritz's closest friends and associates on the staff, along with Margrit and their daughters. A large picture of a youthful Zwicky, in 1931, on Mount Wilson, was displayed prominently on a stand. The institution was represented by Jesse Greenstein, the head of Caltech's physics and astronomy department and a frequent combatant in Zwicky's many battles with grey thinkers. Whatever his private feelings, Greenstein was enough of a diplomat, and enough of a historian of science, to realize that the occasion required gravity and solemnity for a man who had blazed so many pathways to the stars.

"Fritz classified scientists into two categories, eagles and low-fliers," he said.[14] A "low-flier like myself recognized clearly that Fritz was the high-flier. He pursued an extraordinary range of personal interests; international charities, city-planning, mountain climbing, new explosives, exploding stars, crystals and dying stars, and especially galaxies. He always saw the universe in his own, original way, he loved the extraordinary objects it contained—he explained them in his own fashion, sometimes wrong but never dull."

Greenstein alluded only briefly to his struggles with Zwicky, "always an administrator-baiter." Referring to supernovae and neutron stars, Greenstein said he carried the idea of gravitational collapse "much further, contemplating 'pygmy stars' (which do not exist) and 'object Hades' (black holes, which probably do exist)." Black hole science was still in its infancy.

Greenstein turned next to Zwicky's interest in galaxies, the compact galaxies that "have become of special importance with the discovery by Sandage, Schmidt and others of the quasars." His inclusion of Sandage's name would no doubt have irritated Zwicky no end if he happened to be watching the proceedings from some celestial remove.

"It is clear that Zwicky's intuition of the importance of implosion-explosion events was a valuable one. In a sense he was a pioneer of high-energy astrophysics."

In conclusion, Greenstein offered one of the most succinct, yet apt, descriptions of Fritz Zwicky, the man, calling him "an extraordinarily live person."

Two of Zwicky's friends also spoke. George McRoberts recalled his time at Aerojet in 1943, when the little rocket company was struggling to meet its first military contracts, and when its rockets tended to blow up on the test stands.

"He was truly an inventive man, unafraid of new ideas," McRoberts said. "In his book, *Everyone a Genius,* he wrote that 'everyone is, in some very particular sense, unique, irreplaceable and incomparable.' This was, in real life, Fritz Zwicky."

Albert S. Wilson, an accomplished astronomer in his own right, said Fritz possessed the singular characteristic of greatness, in that "few people were merely indifferent to him. . . . Those who see further and deeper are not universally admired."[15]

Zwicky was particularly eager to unhorse the pompous. "He felt that all professors and executives should stay in touch with reality by periodically cleaning the washrooms." Zwicky put it more colorfully, but the listeners knew what Wilson was getting at.

"It would please Zwicky to say that, 'That bastard Chairman Mao stole this aspect of the cultural revolution' from him."

With Fritz Zwicky's passing, Wilson said, "the world becomes more homogenized and more mediocre."

The newspapers that had followed Zwicky's exploits, whether climbing a mountain or imagining how we might rearrange the solar system, took less notice. His hometown *Los Angeles Times* carried only an unimpressive obituary on an inside page.

In contrast, the astronomer Cecilia Payne-Gaposchkin penned a lengthy tribute, which was extraordinary since Zwicky's critics liked to argue that he had been unchivalrous in assessing her work. "Fritz Zwicky was one of the few survivors—and an unforgettable example—of the scientific individualism of earlier years," she wrote.[16]

After ticking off the list of his discoveries and interests, she noted, "nothing in the universe was foreign to his interest, nothing excluded from his study . . . Fritz Zwicky was one of the last of the scientific individualists, a breed that is dying out in an age of teamwork. Aggressively original, outspoken to the point of abrasion, he seemed to his contemporaries stubbornly opinionated. . . . Looking back on his rugged determination and his slightly Renaissance flavor, one is reminded of Tycho Brahe: opinionated, combative, a superb observer and a very human person. For Fritz Zwicky was one of the kindest of men, with a deep concern for humanity."

Despite this praise, it would be Fritz Zwicky's enemies who would have the last word in the next decades. In the scathing assessments by Sandage, Greenstein, and others, Zwicky was transformed into either a cartoonish bumpkin or an agent of pure malice. These denunciations had their apotheosis in Bill Bryson's 2003 book *A Short History of Nearly Everything,* in which he said colleagues considered Zwicky an "irritating buffoon" who "didn't seem outstandingly bright." Only in recent times has this characterization begun to change, as the power of the discoveries endures, while the sniping of the critics, many of whom have died, has faded into history.[17]

In spite of the tributes in America, Zwicky's widow did not find the same eagerness to honor its native son when she approached the authorities in Switzerland with a request to inter his ashes in his hometown of Glarus. Land was at a premium in their small country, she was told. And besides, he had lived abroad most of his adult life. The fact that Fritz Zwicky had given up a lucrative career in the defense industry in the United States rather than surrender his Swiss citizenship, that he had fallen under suspicion as a Communist sympathizer, mattered not at all. Margrit could not help but wonder if the same bureaucratic thinking that frustrated her husband so much in life wasn't playing out after his death. There was also the possibility that the locals found it impossible to overlook the occasional broadsides that Zwicky aimed at the Swiss over the years.

In the end, Margrit was allowed to inter his remains in the small cemetery at Mollis, just a few miles down the highway from Glarus. He shared the patch of earth next to an ancient church with the heroes of the 700-year-old battle of Näfels.

In 1998, on the occasion of Zwicky's hundredth birthday, a wooden monument was erected nearby, resembling the 18-inch Schmidt telescope with which Zwicky had made so many discoveries. Aimed at the heavens, it was a clear symbol and a reminder that the man whose ashes were interred there had aimed higher and dared more than most men.

For some time after his death, Margrit was lost. She had no interest in marrying again. She stayed as busy as she could, taking a job in a convalescent home. "The work is hard and the pay lousy," she wrote to Margritli. Just two dollars an hour. "But at least I have something worthwhile to do."[18]

Two years after her husband's death, she was still living in the same house on Oakdale that Fritz bought for her three decades before. "Last night was Halloween," she wrote. "I had not too many beggars and closed shop at 8:30."[19]

The presidential election was coming. She wondered how it would turn out. Jimmy Carter was opposing Jerry Ford, the man who replaced

Mollis mit Glärnischkette

The town of Mollis, in the Glarus Valley, where Fritz Zwicky's remains were interred, in the same small cemetery as the heroes of the ancient Battle of Näfels. "Nothing in the universe was foreign to his interest, nothing excluded from his study," wrote the astronomer Cecilia Payne-Gaposchkin.

Richard Nixon after his resignation. "I guess it does not matter much which jackass gets in, one is as bad as the other," she added bitterly. "Heaven help this country."[20]

For many years afterward, her daughters tried to convince their mother to move back to Switzerland. She always said it was too much bother. Finally, they packed her up and moved her. She died soon after, at 83, in 2012.

So what was the truth about Fritz Zwicky? Everyone has his own truth, even in science. Most people are willing to collaborate their way to truth, altering their opinions to win converts and supporters. Zwicky could not do that. And he paid for it. Had he been more willing to compromise and see the wisdom of others' views, he would likely have won acclaim from his colleagues. He could have continued working with Baade and shared in the other man's discoveries of stellar populations. He could certainly have found an ally in Hubble, who was an admirer in the early days, when the theory of tired light still seemed credible. When the younger generation came along, he could have welcomed them, instead of retreating into his circumscribed world, relying only on his own insights and investigations at his Schmidt telescope. Like Chandrasekhar and Hubble, he might have had a space telescope named for him.[21] But if he had done all those things, had gone along to get along, he wouldn't have been Fritz Zwicky. And the real Zwicky turned out to be very necessary during those early days, when Hale's giant telescopes revealed the awesome secrets of the universe, when the world needed someone with imagination and courage to suggest that the undiscovered country of space was more lawless and violent than anyone could imagine.

That same spirit was needed in World War II, when the US Navy's efforts in the Pacific were being held back by the difficulty of launching planes from the short-decked aircraft carriers. To wish he were something other than what he was diminishes the great things he did, things that nobody else in his generation had the imagination and daring to attempt. If he was ungentlemanly, if he could be difficult to deal with, he was also truly and singularly unique. It was entirely fitting that the star by which he steered his life was an explosive one.

Fritz Zwicky died less than six months after his middle daughter's marriage. As he walked Fran down the castle path in Spiez on the day of her marriage, Zwicky softly sang an old Swiss military air.[22] With the lake twinkling below, he sang the third verse of the *Beresina,* a battlefield tune from the nineteenth century. "Have faith, have faith, dear brothers, forget your worries and your woes, tomorrow the sun will once again rise in friendly skies."

Fran was not surprised at the martial sentiment. She knew this was one of her father's favorite songs. She also knew that for her father, life was to be grappled with, a constant struggle to wrest civilization from the clutches of disorder and villainy. He was telling her that she must not fear taking this important step, or any other. Her father and the family would always be at her side. And tomorrow, the sun would indeed rise again.

NOTES

ACKNOWLEDGMENTS

ILLUSTRATION CREDITS

INDEX

NOTES

1. Noted Young Men of Science

About the endnotes: Fritz Zwicky wrote, received, and saved many hundreds of letters to and from friends, colleagues, fans, enemies, and family. Many letters reside in the Fritz Zwicky Foundation Archive in Glarus, Switzerland. Though I read and made copies of many, I am confident there is much more to be discovered by future researchers. I also received copies of letters from Alfred Stöckli, Zwicky's Swiss biographer, and from Zwicky's daughters, Margrit (Margritli) Zwicky and Franziska Zwicky-Pfenninger. In some cases, excerpts from those letters contained in this book are identified by source. To avoid numbing the reader I have not identified each and every source in these notes. In cases where the sources are not identified, the reader may assume it is one of the four collections listed here. Further, the details of Zwicky's family life come from a series of interviews I conducted over a period of months with Margritli and Franziska in Pasadena, Glarus, and Mettmenstetten. They kindly supplied additional information by email later. In the book, again for clarity and simplicity, I have sometimes avoided sourcing the details of each interview's location and time. Except where otherwise identified in the text, the quotations from the daughters come from those interviews and emails.

1. "Society History," American Physical Society website, aps.org.
2. Leonard B. Loeb, "Proceedings of the American Physical Society, Minutes of the Stanford Meeting, December 15–16, 1933," *Physical Review* 45, no. 2 (January 15, 1934): 130–139.
3. Loeb, "Proceedings of the American Physical Society, Minutes of the Stanford Meeting," 136.
4. Loeb, "Proceedings of the American Physical Society, Minutes of the Stanford Meeting," 137.
5. R. M. Langer, "Fast New World," *Collier's Weekly*, July 6, 1940, 18–19; 18.

6. Loeb, "Proceedings of the American Physical Society, Minutes of the Stanford Meeting," 138.

7. A nova is an explosion on the surface of a white dwarf star; it does not destroy the star. Ordinary novae are 10,000 times dimmer than supernovae. Zwicky later credited Lundmark's "incredible foresight and imagination" in making the distinction between the types of exploding stars.

8. Robert Kirshner, *The Extravagant Universe* (Princeton University Press, 2002), 32.

9. This was later proven to be true only for smaller galaxies, not big galaxies like the Milky Way. Today, we know supernovae can be as bright as 4 billion suns.

10. Loeb, "Proceedings of the American Physical Society, Minutes of the Stanford Meeting," 138.

11. W. Baade and F. Zwicky, "Cosmic Rays from Super-Novae," *Proceedings of the National Academy of Sciences* 20, no. 5 (May 1934): 259–263; 260.

12. Baade and Zwicky, "Cosmic Rays from Super-Novae," 261. Scientists now know that because cosmic rays bounce around the galaxy due to the Milky Way's magnetic field, tracing any cosmic ray to its source is just as impossible as it was in Zwicky's day. The very highest energy ones, which can come from outside the galaxy, are the exception.

13. Baade and Zwicky, "Cosmic Rays from Super-Novae," 263.

14. The term may be recent, but the idea of nonluminous bodies—where gravity is so strong it prevents light from escaping—can be traced to John Michell and Pierre-Simon Laplace in 1784 and 1796. Michell envisioned a process in which the gravity could be so strong that "all light emitted from such a body would be made to return towards it." Colin Montgomery, Wayne Orchiston, and I. B. Whittingham, "Michell, LaPlace and the Origin of the Black Hole Concept," *Journal of Astronomical History and Heritage* 12, no. 2 (January 2009): 90–96.

15. W. Baade and F. Zwicky, "Remarks on Super-Novae and Cosmic Rays," *Physical Review* 46, no. 1 (July 1, 1934): 76–77; 77.

16. Letter to Rösli Streiff, February 6, 1934.

17. We now know supernovae can be caused both by nuclear reactions and by gravitational collapse. Though Zwicky later called supernovae "atom bomb stars," the type he and Baade proposed—now called a Type II supernova—is produced by gravitational collapse. The true atom bomb star, a Type Ia supernova, involves two stars. In that process, a star known as a white dwarf feeds off its companion until it reaches a critical mass and undergoes a thermonuclear explosion.

18. Beatrice Muriel Hill Tinsley is credited with this prediction. She was an innovator in modeling the evolution of star systems, producing many papers that are still cited. Her career as an astrophysicist was cut short by disease in 1981, at the age of 40.

19. Baade and Zwicky, "Cosmic Rays from Super-Novae," 262.

20. Kip S. Thorne, *Black Holes and Time Warps: Einstein's Outrageous Legacy* (Norton, 1994), 174.

21. F. Zwicky, "Introduction," in *Catalogue of Compact Galaxies & of Post-Eruptive Galaxies* (1971), xv.

22. "Star Suicide," *Time,* July 23, 1934, 50.

23. Dan Noyes, "Cosmic Rays Discovered 100 Years Ago," CERN, August 7, 2012, https://home.cern/news/news/physics/cosmic-rays-discovered-100-years-ago-0.

24. "Be Scientific with OL' DOC DABBLE," *Los Angeles Times,* January 19, 1934.

25. "They Stand Out from the Crowd," *Literary Digest,* August 25, 1934, 13.

26. "Star of Bethlehem Continues to Baffle World Astronomers," *Washington Evening Star,* December 25, 1933.

27. "Stars Are Not 'Eternal, Unchangeable,' Asserts Pasadena Astrophysicist," *Pasadena Star-News,* March 9, 1935.

28. Letter to Zwicky from Streiff, April 15, 1935.

2. Glarus Thrust

1. F. Zwicky, *Everyone a Genius,* English version, unpublished, 10, Zwicky Archive of the Fritz Zwicky Foundation, in the Landesbiblioteck, the main library in Glarus, Switzerland.

2. Josef Schwitter and Urs Heer, *Glarnerland: A Short Portrait* (Verlag Baeschlin, 2000), 24–26.

3. Max Eastman, "The Oasis of Europe," *Readers Digest,* March 1952, 77.

4. Schwitter and Heer, *Glarnerland,* 40.

5. Alfred Stöckli and Roland Müller, *Fritz Zwicky: An Extraordinary Astrophysicist,* trans. Ian Gordon (Cambridge Scientific Publishers, 2011), 5.

6. Unless otherwise specified, Zwicky's often pithy quotations in this chapter were taken from Zwicky, *Jeder ein Genie* [Everyone a Genius] (Ex Libris, 1972) or Stöckli and Müller, *Fritz Zwicky: An Extraordinary Astrophysicist.*

7. Stöckli and Müller, *Fritz Zwicky: An Extraordinary Astrophysicist,* 4.

8. Zwicky, *Jeder ein Genie.*

9. Interview with R. Cargill Hall, historian, Jet Propulsion Laboratory, May 17, 1971, 3.

10. Stöckli and Müller, *Fritz Zwicky: An Extraordinary Astrophysicist,* 7.

11. Stöckli and Müller, *Fritz Zwicky: An Extraordinary Astrophysicist,* 11.

12. At the time he wrote the letter, the question of how much material existed outside the galaxies was a pressing one. Zwicky's insistence that there was a lot of stuff on the frontiers of space was prescient.

13. Letter from Zwicky to Reichstein, October 8, 1952.

14. Stöckli and Müller, *Fritz Zwicky: An Extraordinary Astrophysicist,* 13.

15. Stöckli and Müller, *Fritz Zwicky: An Extraordinary Astrophysicist,* 12.

16. Seminar on morphology at Caltech, undated, 6, Zwicky archive, Glarus.

17. Stöckli and Müller, *Fritz Zwicky: An Extraordinary Astrophysicist,* 13.

18. Stöckli and Müller, *Fritz Zwicky: An Extraordinary Astrophysicist,* 21.

3. The Bigger and Better Elephant

1. Alfred Stöckli and Roland Müller, *Fritz Zwicky: An Extraordinary Astrophysicist,* trans. Ian Gordon (Cambridge Scientific Publishers, 2011), 15.

2. Fritz Zwicky, "Achievements of Einstein," *Pasadena Star-News,* undated, Zwicky Archive of the Fritz Zwicky Foundation, in the Landesbiblioteck, the main library in Glarus, Switzerland.

3. Alfred E. Senn, *The Russian Revolution in Switzerland* (University of Wisconsin Press, 1971), 150.

4. Alexander Solzhenitsyn, *Lenin in Zurich* (Farrar, Strauss & Giroux, 1976), 41.

5. Senn, *The Russian Revolution in Switzerland,* 150.

6. Senn, *The Russian Revolution in Switzerland,* 216.

7. Seminar on Morphology at Aerojet, May 17, 1957, 5, Zwicky archive, Glarus.

8. Seminar on Morphology at Aerojet, 5.

9. Seminar on Morphology at Aerojet, 6.

10. Seminar on Morphology at Aerojet, 6.

11. Seminar on Morphology at Aerojet, 7.

12. Seminar on Morphology at Aerojet, 7.

13. F. Zwicky, "Free World Agents of Democracy," *Engineering and Science* 13, no. 2 (November 1949): 10–14.

14. Stöckli and Müller, *Fritz Zwicky: An Extraordinary Astrophysicist,* 16.

15. Theodore von Kármán, with Lee Edson, *The Wind and Beyond* (Little, Brown and Co., 1967), 55.

16. Interview with R. Cargill Hall, historian, Jet Propulsion Laboratory, May 17, 1971, 2.

17. Interview with R. Cargill Hall, May 17, 1971, 2.

18. Allan Sandage, *Centennial History of the Carnegie Institution of Washington, Volume I* (Cambridge University Press, 2004), 12.

19. Story recounted by Zwicky several years after his arrival, reported by the *Los Angeles Times,* September 28, 1931, part II, 3.

20. The charge was 1.602×10^{-19} coulombs; 7.49×10^{19} electrons flow through a light bulb every minute to keep it lit.

21. Though some were awarded for work done before the honoree arrived on campus.

22. Fritz Zwicky, *American National Biography,* Vol. 24 (Oxford University Press, 1999), 269.

23. Stöckli and Müller, *Fritz Zwicky: An Extraordinary Astrophysicist,* 25.

24. From a collection of letters written between Rösli Streiff and Fritz Zwicky, from 1930 until 1948. This was from February 15, 1931.

25. Letter to Rösli, February 15, 1931.

26. Apparently this was Big Pines, near Wrightwood, where people were jumping over 200 feet. This letter was written February 15, 1931.

27. Letter to Rösli, January 1932.

28. Letter to Rösli, September 27, 1931.

29. F. Zwicky, "On the Physics of Crystals, Part I," *Reviews of Modern Physics* (July 1934): 193–208.

30. Letter to Rösli, February 15, 1931.

31. "Teachers Hear Physicist," *Los Angeles Times,* December 15, 1932.

32. "Teachers Hear Physicist," *Los Angeles Times.*

33. Zwicky had a raconteur's wit and facility for bending the truth in the service of a good story. A story he enjoyed telling over the years concerned his friend, the physicist Paul Epstein. While driving with friends from Caltech one day, Epstein (or Eppy, in Zwicky's telling) lost his glasses. Zwicky offered to drive, but Epstein refused, saying he would rely on mathematics and physics to guide his ancient Buick to its destination. When the car plunged into an orange grove instead, Epstein got out and gravely addressed his companions: "Gentlemen, gentlemen, not since I was in gymnasium have I made such a silly miscalculation." Zwicky again volunteered to take the wheel and steer them out of the

grove. Epstein again refused, saying he would make new and better calculations to navigate through the trees. Everyone got back in, and Eppy was as good as his word, gliding past "the first orange tree, the second, the third, the fourth, not touching a leaf, all by mathematics. Eppy sees nothing and we come on the highway precisely at the right angle. Marvelous mathematician Eppy," pronounced Zwicky. There was only one problem: he failed to see another car arrive at the same spot on the highway at the same moment. "And what a hell of a mess that was." This story was recounted by *Los Angeles Times* columnist Jack Smith in 1980, under the heading, "Which Way Did They Go?," January 6, 1980, 83.

4. Quantum Steak and Matrix Salad

1. Lorentz's theories of the electron and electromagnetics assumed the existence of a stationary ether, which Einstein discarded. Henri Poincaré wrote, "Whether the ether exists or not matters little; let us leave that to the metaphysicians." Henri Poincaré, *Science and Hypothesis* (Walter Scott Publishing Co., 1905), 211.

2. In 1927, UCLA, previously known as UC's Southern Branch, was located on Vermont Avenue. It opened its Westwood campus in 1929.

3. Alfred Stöckli and Roland Müller, *Fritz Zwicky: An Extraordinary Astrophysicist,* trans. Ian Gordon (Cambridge Scientific Publishers, 2011), 27.

4. Letter from Fridolin to his son, June 29, 1927, trans. Margritli Zwicky.

5. Letter from Fridolin to his son, June 29, 1927, trans. Margritli Zwicky.

6. Stöckli and Müller, *Fritz Zwicky,* 6.

7. An opinion shared by his father, who replied in a letter in May 1927 that the "clipping you sent show us the mountains in America are just hills."

8. "Mount Whitney Climbers' Goal, Educators Will Defy Perils Again," *Los Angeles Times,* April 3, 1927, 1.

9. Two years later, however, Zwicky succeeded in the attempt, using skis to get through the deep snow.

10. "Caltech Forms Rescue Party, Two Physicists Snowbound Five Days in Hills," *Los Angeles Times,* November 30, 1931, 15.

11. *Scientific American,* March 1932. The article on the first page of the edition also said Zwicky's name "has figured prominently in discussions of cosmology among the most noted men of science."

12. Six-page, typed account in the Zwicky Archive of the Fritz Zwicky Foundation, in the Landesbiblioteck, the main library in Glarus, Switzerland. Brunner wrote that the two climbs "were equal to a dozen of easier High Mountain climbs."

13. Stöckli and Müller, *Fritz Zwicky,* 21.

14. Bill Bryson, *A Short History of Nearly Everything* (Broadway Books, 2003), 31.

15. Using mountains as a metaphor for the trials of science was not unique to Zwicky. The astronomer Cecilia Payne-Gaposchkin advised those thinking of entering the field not to expect fame or money. "Undertake it only if nothing else will satisfy you," she wrote, "for nothing is probably what you will receive. Your reward will be the widening of the horizon as you climb. And if you achieve that reward you will ask no other." Doug West, "Introduction," in *The Astronomer Cecilia Payne-Gaposchkin: A Short Biography* (Doug West, 2015), vii.

5. The Expanding Universe and Tired Light

1. Richard Berendzen, Richard Hart, and Daniel Seeley, *Man Discovers the Galaxies* (Columbia University Press, 1984), 11.

2. Allan Sandage, *Centennial History of the Carnegie Institution of Washington, Volume 1, The Mount Wilson Observatory* (Cambridge University Press, 2005), 213.

3. P. C. van der Kruit, "Lessons from the Milky Way: The Kapteyn Universe," in *Lessons from the Local Group,* ed. K. C. Freeman et al. (Springer, 2015), 24. The idea was unveiled to the public at the International Congress of Arts and Sciences in St. Louis, 1904. The program featured sessions on metaphysics and "mental science." A chance meeting with Hale at the exhibition proved important for both men later on.

4. Isaac Newton, who attended Trinity College at Cambridge University with the intent of becoming a minister, was not so fulsome. For him, the universe wasn't simply a show. The purpose of understanding it, he believed, was that it would bring us closer to the Creator.

5. Sandage, *Centennial History,* 225.

6. Virginia Trimble, "The 1920 Shapley-Curtis Discussion: Background, Issues, and Aftermath," *Publications of the Astronomy Society of the Pacific* 107, no. 718 (December 1995): 1133–1144.

7. J. D. Fern, "The Great Debate," *American Scientist* 83 (1995): 410.

8. Berendzen, Hart, and Seeley, *Man Discovers the Galaxies,* 45.

9. Hubble discovered even earlier that little NGC 6822, known as Barnard's Galaxy, 1.5 million light-years away, was a separate galaxy. But Andromeda, big and bright, got all the glory. At the time, Hubble estimated a distance of 1 million light-years. With better measurements, we now know Andromeda is 2.6 million light-years away.

10. Corey S. Powell, "January 1, 1925: The Day We Discovered the Universe," *Discover Magazine,* January 2, 2017.

11. F. Zwicky, "On the Thermodynamic Equilibrium in the Universe," *Proceedings of the National Academy of Sciences* 14, no. 7 (1928): 592–597.

12. Alfred Stöckli and Roland Müller, *Fritz Zwicky: An Extraordinary Astrophysicist,* trans. Ian Gordon (Cambridge Scientific Publishers, 2011), 48.

13. Letter to Rösli on February 2, 1933, trans. Margritli Zwicky.

14. He wrote that the centers of spiral galaxies were places where new material might be streaming into our universe from somewhere else.

15. *Los Angeles Times,* April 29, 1931, 28.

16. Ransome Sutton, "Exploding Universe Seen By Savant," *Los Angeles Times,* May 10, 1931, 19–20.

17. F. Zwicky, "How Far Do Cosmic Rays Travel?" *Physical Review* 43, no. 2 (1933): 147–148.

18. David O. Woodbury, "Behold the Universe," *Collier's* (May 7, 1949), 64.

19. Richard Preston, *First Light: The Search for the Edge of the Universe* (Random House, 1987), 44.

20. Marianne Freiberger, "Schrodinger's Equation—What Is It?" *+Plus* (August 2, 2012).

21. Stöckli and Müller, *Fritz Zwicky: An Extraordinary Astrophysicist,* 31.

22. Lemaître produced, in 1927, a prescient paper that many think should allow him at least to share credit for discovering the expanding universe: Georges Lemaître, "A Homogeneous Universe of Constant Mass and Increasing Radius Accounting for the Radial Velocity of Extra-Galactic Nebulae," *Annales de la Société scientifiques de Bruxelles* A47 (1927): 49.

23. Laird Thompson, Astronomy Department, University of Illinois, "Vesto Slipher and the First Galaxy Redshifts," arxiv.org. (2011).

24. Sandage, *Centennial History,* 503.

25. Interview with R. Cargill Hall, historian, Jet Propulsion Laboratory, May 17, 1971.

26. Lizzie Buchen, "May 29, 1919: A Major Eclipse, Relatively Speaking," *Wired,* May 29, 2009.

27. F. Zwicky, "The Redshift of Extragalactic Nebulae," *Helvetica Physica Acta* 6 (1933): 110–127.

28. Barry Parker, "Discovery of the Expanding Universe," *Sky & Telescope* (September 1986): 229.

29. He was so cautious that he used the term "apparent velocity" to describe his measurements. In a 1931 note to the brilliant mathematician Willem de Sitter, quoted by the physicist and author Robert Kirshner, Hubble writes that he wasn't yet convinced his measurements proved the universe was expanding. "The interpretation" of his results, he said meekly, "should be left to you and the very few others who are competent to discuss the matter with authority." Robert P. Kirshner, *The Extravagant Universe* (Princeton University Press, 2002), 91.

30. Edwin Hubble, *The Realm of the Nebulae* (Oxford University Press, 1936).

31. Hubble also wrote, "Although no other plausible explanation of red-shifts has been found, interpretation as velocity shifts may be considered as a theory still to be tested." Hubble, *The Realm of the Nebulae,* 34.

32. Zwicky, "The Redshift of Extragalactic Nebulae."

33. Yet, in recent years, scientists have started to take seriously the idea that the Big Bang was not perhaps the start of everything. "What came before the Big Bang is no longer a silly question," says Virginia Trimble. Personal correspondence, 2018.

34. "Though he used redshift measurements years later in his own work, he qualified them as "indicative velocities." This quote is from a biographical sketch in the Zwicky archive by his friend and associate, Paul Wild, Astronomical Institute, University of Berne, 3.

35. Interview with Ellis at Caltech, 2014.

36. Interview with Djorgovski at Caltech, 2015.

37. Jack Smith column, "Caltech Life: Send in the Clowns," *Los Angeles Times,* May 19, 1981.

38. Notes in Zwicky's hand from the Zwicky Archive of the Fritz Zwicky Foundation, in the Landesbiblioteck, the main library in Glarus, Switzerland, undated.

39. Zwicky, "The Redshift of Extragalactic Nebulae."

40. Okay, not quite no one. In 1930, Knut Lundmark, whose early work on exploding stars Zwicky admired, had used the term *Dunkle Materie* to describe unseen material in galaxies. But his idea received even less attention than Zwicky's and remained for-

gotten for decades, until Vera Rubin's measurements in the 1970s reawakened interest in the subject and in Fritz Zwicky. Lundmark may not have appreciated his own insight. He appeared to think the hidden material was just stuff we couldn't yet see, not some entirely new category of matter.

41. In an October 1937 paper for the *Astrophysical Journal,* "On the Masses of Nebulae and of Clusters of Nebulae" (p. 218), Zwicky did make some initial guesses about the nature of dark matter. To nail down the masses of galaxy clusters, he said, "We must know how much dark matter is incorporated in nebulae in the form of cool and cold stars, macroscopic and microscopic solid bodies, and gases."

42. Interview with Sunil Golwala, Caltech, 2014.

43. Ransome Sutton, "Caltech Scientists Plan Reception of Einstein," *Los Angeles Times,* December 28, 1930, 1.

44. Judith Goodstein, "Albert Einstein in California," *Engineering and Science* 42, no. 5 (1979): 17–19.

45. Stöckli and Müller, *Fritz Zwicky, An Extraordinary Astrophysicist,* 42.

46. Stöckli and Müller, *Fritz Zwicky, An Extraordinary Astrophysicist,* 43.

47. Californios are those who can trace their ancestry to the Spanish land grant recipients of an earlier age.

48. Zwicky translated a revealing poem of Einstein's from the German, which he saved and is held in his archive. The last stanza went: "Often amidst this happy foment," meaning his fame, "Do I ask myself in lucid moment / Am I nuts myself perhaps / Or are all the others saps."

The Skeleton of the Universe

1. There is a lot of debate about how much credit should be given to Rubin, who died on Christmas Day, 2016, at age 88. Albert Bosma, Jean Einasto, J. Ostriker, and E. Saar all made important contributions to the understanding of dark matter as early as 1974. And "damnit, Fritz's analysis of the dynamics of the Coma Cluster and of Virgo decades before were just as convincing," says James Gunn, the Emeritus Eugene Higgins Professor of Astrophysical Sciences at Princeton University.

2. There remain some physicists and astronomers who think the best explanation of the data is some form of gravity outside Einsteinian relativity.

3. Interview, May 29, 2015.

4. Fermi also gave his name to the Fermion, a family of particles with spin ½. The class includes particles most people know—electrons, neutrons, and protons.

6. New Alliances, New Physics

1. Letter to Rösli, November 26, 1932.

2. John S. McGroarty, ed., *History of Los Angeles County* (American Historical Society, 1923), II:346–347.

3. It was called the El Arco Syndicate, in the northern district of what was called Lower California.

4. "Favors New Jury Law," *Los Angeles Times,* March 23, 1923, 1.

5. "Pasadena Girl Weds Scientist," *Pasadena Star-News,* March 26, 1932.

6. Alfred Stöckli and Roland Müller, *Fritz Zwicky: An Extraordinary Astrophysicist,* trans. Ian Gordon (Cambridge Scientific Publishers, 2011), 46.

7. Letter to Rösli Streiff, April 12, 1932.

8. "Alaskan Trip Enjoyed," *Pasadena Star-News,* September 18, 1934, 9.

9. Letter to Rösli, February 2, 1933.

10. Letter to Rösli Streiff, February 2, 1933, trans. Margritli Zwicky.

11. Donald E. Osterbrock, *Walter Baade, A Life in Astrophysics* (Princeton University Press, 2001), 6.

12. The observatory's researchers were not alone in this fantasy. Percival Lowell, who founded the Lowell Observatory, said the canals were built by a dying civilization tapping the polar ice caps for water.

13. Osterbrock, *Walter Baade,* 44.

14. *Daily Capital Journal,* December 9, 1933, 7.

15. It's now known that neutrons in a collapsing neutron star are first made in the center of the star, not at the surface. In 1937, J. Robert Oppenheimer and George Volkoff would develop the mathematical theory of neutron stars, something Zwicky did not do. Had he done that, he might have arrived early on at the concept of black holes. But Zwicky had already moved on.

16. W. Baade and F. Zwicky, "Remarks on Super-Novae and Cosmic Rays," *Physical Review* 46, no. 1 (July 1, 1934): 76–77.

17. F. Zwicky, "On a New Type of Reasoning and Some of Its Possible Consequences," *Physical Review* 43, no. 12 (June 15, 1933): 1031–1033. And F. Zwicky, "Intrinsic Variability of the So-Called Fundamental Physical Constants," letter to *Physical Review* 53 (1938): 315.

18. Letter to the editor, *Philosophy of Science,* 487, in response to Zwicky's article in *Philosophy of Science* 1, no. 1 (July 1934): 353–358.

19. Ransome Sutton, "Fifty Years from Now," *Los Angeles Times Magazine,* July 30, 1933, 3.

20. Fritz Zwicky, *Novae Become Supernovae* (Koenig, T. Astronomical Society of the Pacific, 2005), 56.

21. Astronomers prefer to use parsecs for distances in space. It is the distance that light travels in 3.26 years, about 20 trillion miles. I prefer to use light-years, as it is easier for laypersons like myself to visualize, though even that is a relative concept when describing the vast wilderness of the stars. The Virgo cluster is around 10 megaparsecs away.

22. Press release, "Three Schmidt Telescopes, 18-Inch, 8-Inch and 48-Inch Apertures," Office of Public Relations, California Institute of Technology, June 19, 1949, 4.

23. Press release, "Three Schmidt Telescopes, 18-Inch, 8-Inch and 48-Inch Apertures," 4.

24. In later years, according to Virginia Trimble, both Zwicky and Baade claimed credit for first recognizing the genius of this insightful, eccentric man.

25. Interview of Fritz Zwicky by R. Cargill Hall, Jet Propulsion Laboratory historian, May 17, 1971, 3.

26. According to an exhibit at Palomar today, the optics were fashioned in the Caltech optical shop. The tube and mounting were made by the Caltech astrophysics machine

shop. What work Zwicky did is not clear, but it's likely he pitched in to help with construction on the 18-foot dome.

27. Most of Zwicky's discoveries of supernovae were made at the 48-inch Samuel Oschin telescope on Palomar. But the 18-inch was special to him, and after he died, the memorial erected to him in Switzerland was a wood replica of the humble telescope he helped build himself.

28. William S. Barton, "Our Expanding Universe," *Los Angeles Times Magazine*, August 23, 1936, 1.

29. Zwicky was assisted in his search by Josef Johnson of Caltech.

30. Letter to Rösli, April 2, 1937, trans. Margritli Zwicky.

31. Twelve-page biographical sketch by R. Cargill Hall, at the time the Jet Propulsion Laboratory historian. This is on page 4.

32. Ironically, according to Robert Kirshner, author of *The Extravagant Universe: Exploding Stars, Dark Energy, and the Accelerating Cosmos* (Princeton University Press, 2002), all of Zwicky's early discoveries proved to be Type Ia supernovae, the kind that doesn't collapse into a neutron star. But at this early point, the DNA of exploding stars had not yet been discovered.

33. Big spiral galaxies like our own are now known to have at least one supernova eruption per century, but Zwicky's early researches tended to focus on smaller galaxies, for which one every thousand years is not such a bad guess.

34. "Suggestions of Obscure Dishwasher May Add to Knowledge of Universe," *Pasadena Star-News*, February 1, 1937, 1, part 2.

35. Letter from Einstein to Mandl, May 12, 1936, EA 17034/35.

36. A. Einstein, *Science* 84 (1936): 506.

37. F. Zwicky, "Nebulae as Gravitational Lenses," *Physical Review* 51, no. 4 (February 15, 1937): 290.

38. This referred to the work of Sinclair Smith, who in 1936 published an analysis of the Virgo cluster of 500 galaxies, arriving at a similar result as Zwicky had done with the Coma cluster.

39. The technique has also finally been used with stars. Even if only one star in a million lines up with another, it is still possible, using today's advanced techniques, to find some alignments by observing millions of stars. The technique was used to search for the halo of dark matter in the Milky Way.

40. The lumpiness left behind proved that dark matter is not affected by electromagnetic forces.

41. Will Jonathan, "The Chief Watchman," *Saturday Review*, July 4, 1964, 43.

42. Letter to Zwicky, August 8, 1939.

43. He also created a popular mnemonic for the classification of star types, which ranged from O, the hottest, to S, at the other end of the scale. As memorized by generations of student astronomers, the phrase went "Oh, Be A Fine Girl, Kiss Me Right Now, Sweet."

44. This is true only for the white dwarf, Type Ia supernova. For that type, a star with 1.4 times the mass of the sun feeds off a companion until it reaches a critical state and explodes. The Type II, Zwicky's and Baade's type, requires a star with at least eight times

the mass of our sun. That type generates an iron core of 1.4 times the mass of the sun, then collapses to form a neutron star (or black hole) with the energy fueling a supernova.

45. F. Zwicky, "On Collapsed Neutron Stars," *Astrophysical Journal* 88 (1938): 522–525.

46. "Stars May Shine Unseen," *Engineering & Science* (1939), 10.

47. Osterbrock, *Walter Baade,* 61.

48. Because his status as an enemy alien meant that he could not work in the war effort, he had the 100-inch Hooker almost to himself during the war years. This contributed materially to his breakthrough discovery of different stellar populations.

49. Osterbrock, *Walter Baade,* 217.

50. Osterbrock, *Walter Baade,* 77.

51. Allan Sandage, *Centennial History of the Carnegie Institution of Washington, Volume 1, The Mount Wilson Observatory* (Cambridge University Press, 2005), 387.

52. Kip S. Thorne, *Black Holes and Time Warps: Einstein's Outrageous Legacy* (W. W. Norton, 1994), 166.

53. Letter to Rösli, December 30, 1945, trans. Margritli Zwicky.

54. Letter to Rösli, February 12, 1946.

55. The photos were of distant nebulae. Their arrangement in space, he said, would take billions of years longer to achieve than the age of the universe. Of course, the age of the universe was about to grow much longer, due mostly to Baade.

56. Poem, unknown source, dated 1939, Zwicky Archive of the Fritz Zwicky Foundation, in the Landesbiblioteck, the main library in Glarus, Switzerland.

7. Tilting at Windmills

1. Letter from Nick Roosevelt to Zwicky and Dorothy, August 21, 1939.

2. Nicholas Roosevelt, *A Front Row Seat* (University of Oklahoma Press, 1953), 226.

3. This and the next several quotes are from a letter from Nick Roosevelt, August 21, 1939.

4. Letter from Fridolin to Fritz, March 24, 1940.

5. Letter to Rösli Streiff, March 11, 1941.

6. R. W. Gerard, "Peace Resolution of the American Association of Scientific Workers," *Science* 91, no. 2373 (June 21, 1940): 596–597.

7. Roosevelt, *A Front Row Seat,* 215–216.

8. Oort was responsible for many other achievements. The Oort cloud beyond Pluto was named for him.

9. Letter to Rösli Streiff, March 11, 1941.

10. F. Zwicky, "Free World Agents of Democracy," *Engineering and Science* 13, no. 2 (November 1949): 12.

11. F. Zwicky, *Morphological Astronomy* (Springer-Verlag, 1957), 267.

12. Letter from Fridolin, March 24, 1940.

13. Letter from Fridolin, March 24, 1940.

14. This and the following quote are from Nick Roosevelt, letter to Baruch, June 14, 1940.

15. Letter from Zwicky, July 17, 1940.

16. Quotes in this and subsequent paragraphs are from a letter to Blackett, November 11, 1940.

17. Letter from Blackett, January 17, 1941.

8. Rocket Man

1. Frank J. Taylor, "Jatos Get 'Em Up," *Saturday Evening Post,* May 19, 1945.
2. Taylor, "Jatos Get 'Em Up."
3. "Navy Patrol Bomber Military Requirements," undated, Zwicky Archive of the Fritz Zwicky Foundation, in the Landesbiblioteck, the main library in Glarus, Switzerland, 1–7.
4. Zwicky archive, Glarus.
5. Theodore von Kármán, with Lee Edson, *The Wind and Beyond* (Little, Brown and Co., 1967), 235.
6. Von Kármán, *The Wind and Beyond,* 235.
7. Von Kármán, *The Wind and Beyond,* 253.
8. Interview with R. Cargill Hall, historian, Jet Propulsion Laboratory, May 17, 1971, 5.
9. Interview with R. Cargill Hall, May 17, 1971, 15.
10. Von Kármán, *The Wind and Beyond,* 6.
11. Aerojet seminar on Morphology, May 17, 1957, 4.
12. F. Zwicky, *Morphological Astronomy* (Springer-Verlag, 1957), 7.
13. Von Kármán, *The Wind and Beyond,* 260.
14. "Squad Of 85 Men Trained to Clean Up Gassed Areas," *Pasadena Post,* June 27, 1943.
15. F. Zwicky, "Activities of the Research Department of the Aerojet Engineering Corporation from September, 1943 to April, 1945," Zwicky archive, Glarus.
16. Andrew G. Haley, "Aerojet Engineering Corporation, Its Background, Objectives and Accomplishments," Aerojet report no. 44, May 21, 1944, 6.
17. Interview with R. Cargill Hall, May 17, 1971.
18. Von Kármán, *The Wind and Beyond,* 264.
19. Haley, "Aerojet Engineering Corporation," 2.
20. Remarks upon Zwicky's death at a memorial at Caltech, March 12, 1974.
21. Von Kármán, *The Wind and Beyond,* 261.
22. Alfred Stöckli and Roland Müller, *Fritz Zwicky: An Extraordinary Astrophysicist,* trans. Ian Gordon (Cambridge Scientific Publishers, 2011), 86.
23. "Eyewitnesses Describe Jet Plant Explosion," *Pasadena Star-News,* August 22, 1946, 1.
24. "Thrust," *Fortune,* September 1946, 141.
25. Von Kármán, *The Wind and Beyond,* 316.
26. Von Kármán, *The Wind and Beyond,* 318.
27. Haley, "Aerojet Engineering Corporation," 21.
28. This and the following quotes are from Taylor, "Jatos Get 'Em Up."

9. Home Fires

1. The divorce was finalized September 12, 1941, in Nevada, two months after Dorothy's mother, Dorothy Stiles Gates, died.
2. Letter to Rösli Streiff, March 11, 1941.
3. Alfred Stöckli and Roland Müller, *Fritz Zwicky: An Extraordinary Astrophysicist,* trans. Ian Gordon (Cambridge Scientific Publishers, 2011), 76.
4. Letter from Fridolin Zwicky, October 1941.

5. Letter to Rösli Streiff, September 19, 1945, trans. Margritli Zwicky.

6. Letter to Rösli Streiff, September 19, 1945, trans. Margritli Zwicky.

7. Virginia Trimble recalled that students in his classical mechanics class invented a student named Hjalmar Sciatti to do homework and take tests. Zwicky eventually caught on to the game, but, enjoying the joke, went along with it. "Anyhow, the class was HARD," Trimble said.

8. Stöckli and Müller, *Fritz Zwicky*, 74.

9. Letter from Robert Millikan, September 6, 1944.

10. Letter to Robert Millikan, August 28, 1944.

11. F. Zwicky, "Reynolds Numbers for Extragalactic Nebulae," *Astrophysical Journal* 43 (1941): 411.

12. Interview with historian R. Cargill Hall, Jet Propulsion Laboratory, May 17, 1971.

13. Letter to Fritz Zwicky from Otto Struve, December 13, 1940.

14. But he also was an imaginative thinker. He was among the minority of astronomers who believed, on the basis of his study of the rotation of stars, that planets harboring intelligent life were ubiquitous in the universe. His idea was rooted in the fact that relatively cool stars, like our sun, rotate more slowly. He thought this was caused by orbiting planets—which might harbor life—slowing the stars' rotation. In fact, all stars start out as rapid rotators and slow over time. Yet Struve's belief in the ubiquity of planets has found vindication in the Kepler spacecraft's discovery of thousands of them around other stars.

15. Letter to Fritz Zwicky from Otto Struve, December 26, 1940.

16. In his book, *Morphological Astronomy* (Springer-Verlag, 1957), Zwicky took a swipe at Edwin Hubble, writing that galaxy clusters "are not relatively rare as Hubble stated in his *Realm of Nebulae*, in 1936." In that book, Hubble had given credit for uncovering the mysterious "missing mass" in galaxies not to Zwicky, but to Sinclair Smith, which no doubt infuriated Zwicky.

17. Handwritten note by Fritz Zwicky on letter from Alexander Pogo, January 23, 1953.

18. Letter to Alexander Pogo, December 14, 1953.

19. Allan Sandage, *Centennial History of the Carnegie Institution of Washington, Volume I, The Mount Wilson Observatory* (Cambridge University Press, 2005), 532.

20. "Sequence of Events which Led to the Suppression of Four of My Scientific Articles," undated, Zwicky Archive of the Fritz Zwicky Foundation, in the Landesbiblioteck, the main library in Glarus, Switzerland.

21. Edwin Hubble, "Problems of Nebular Research," *Scientific Monthly* 51 (November 1941): 391–408.

22. Commenting on Zwicky's insight in the book, *The Road to Galaxy Formation* (Springer-Verlag, 2007), author William C. Keel wrote (p. 22) that "it is striking to note how often in the history of ideas about galaxies we find that Zwicky got there first."

23. Fritz Zwicky, "Introduction," in *Catalogue of Selective Compact and of Post-Eruptive Galaxies*, Volume 6 (Guemligen, Switzerland, 1971), 8, Zwicky archive, Glarus.

24. Letter from Fridolin Zwicky, January 10, 1944.

25. Letter to Fridolin Zwicky, October 17, 1942.

26. Letter to Rudi Zwicky, January 15, 1951.

27. Letter from Rudi Zwicky, December 30, 1944.

28. Letter to Leonie Zwicky-Staneff, May 23, 1947.

29. Letter from Elfriede and Rudi, February 26, 1947.

30. Transfer notarized in Los Angeles County, March 14, 1972.

31. Letter from Rudi Zwicky, February 11, 1945.

10. Secret Missions, Finding von Braun

1. Theodore von Kármán, with Lee Edson, *The Wind and Beyond* (Little, Brown and Co., 1967), 268.

2. Zwicky actually made two trips, one in the weeks just after the war and the other in 1946.

3. Fritz Zwicky, *Report on Certain Phases of War Research in Germany,* Zwicky Archive of the Fritz Zwicky Foundation, in the Landesbiblioteck, the main library in Glarus, Switzerland.

4. Daniel Lang, "A Reporter at Large, a Romantic Urge" (*New Yorker,* April 21, 1951), 68–85, quotations from pages 78–79.

5. Lang, "A Reporter at Large, a Romantic Urge," 79.

6. F. Zwicky, *Morphology of Propulsive Power* (Society for Morphological Research, 1962), 135.

7. Zwicky, *Morphology of Propulsive Power,* 139.

8. Hahn did the experiment, but the physicist Lise Meitner explained the process. It was her nephew, Otto Frisch, who came up with the term "fission," adapting the name used in biology for cell division.

9. Von Kármán, *The Wind and Beyond,* 283.

10. McGeorge Bundy, "Hitler and the Bomb," *New York Times,* November 13, 1988.

11. Zwicky, *Morphology of Propulsive Power,* 136.

12. Von Kármán, *The Wind and Beyond,* 273.

13. Von Kármán, *The Wind and Beyond,* 274.

14. Von Kármán, *The Wind and Beyond,* 277.

15. Von Kármán, *The Wind and Beyond,* 279.

16. Letter from Robert Millikan to Caltech associates, August 27, 1945, talk scheduled for September 13, 1945.

17. Letter from U.S. Naval Technical Mission in Europe, June 22, 1945.

18. He proposed, in a September letter to von Kármán, that this time he should be given the rank of brigadier general, instead of colonel, "which everybody has." The request was not honored.

19. Notes from the Zwicky archive, Glarus. Estimates of the dead by other sources vary widely, but a 1946 report by the Manhattan Engineer District said 66,000 were killed, about 45,000 of them on the first day, which is close to Zwicky's figure.

20. F. Zwicky, *Morphological Astronomy* (Springer-Verlag, 1957), 267.

21. Zwicky, *Morphological Astronomy,* 268.

22. Account of Oppenheimer visit to Caltech, "Distinguished Visitor II," *Engineering & Science* (February 1950), 23, Zwicky archive, Glarus.

23. Alfred Stöckli and Roland Müller, *Fritz Zwicky: An Extraordinary Astrophysicist,* trans. Ian Gordon (Cambridge Scientific Publishers, 2011), 98.

24. Alexander de Seversky, "Atomic Bomb Hysteria," *Reader's Digest* (February 1946), 121–126.

25. "Survival Is Your Personal Business," Aerojet-General pamphlet, undated, Zwicky archive, Glarus.

26. Walter Libby, "You Can Survive Atomic Attack," *Pasadena Star-News,* November 1961.

27. Letter to von Kármán, January 29, 1946.

28. Fritz Zwicky, 1946 diary entry, Zwicky archive, Glarus.

29. Zwicky, *Morphological Astronomy,* 264.

30. Zwicky, *Morphological Astronomy,* 269.

31. Zwicky, *Morphological Astronomy,* 265.

32. Letter to Rösli Streiff, August 3, 1945, from the Hotel Statler.

33. Stöckli and Müller, *Fritz Zwicky,* 77.

34. Letter to Rösli Streiff, February 12, 1946.

11. The March into Space

1. It later became Wright-Patterson Air Force Base. Letter from John O'Mara, March 5, 1946.

2. "RAAF Captures Flying Saucer on Ranch in Roswell Region," July 8, 1947, 1.

3. "Underwater Jet," *Time* (March 14, 1949), 88.

4. "Underwater Jet," 88.

5. "Rockets in the Desert," *Army Ordnance* (March–April 1947).

6. F. Zwicky, *Morphology of Propulsive Power* (Society for Morphological Research, 1962), 146.

7. James E. Bassett, "Rocket Test May Give Key to Space Travel," *Los Angeles Times* (December 13, 1946), 2.

8. F. Zwicky, "The First Shots into Interplanetary Space," *Engineering and Science* (Caltech, January 1958), 20–23.

9. "Man-Made METEORITE to Be Launched," *Science Illustrated* (September 1946), 56.

10. Zwicky, *Morphology of Propulsive Power,* 146.

11. Alfred Stöckli and Roland Müller, *Fritz Zwicky, An Extraordinary Astrophysicist,* trans. Ian Gordon (Cambridge Scientific Publishers, 2011), 104.

12. Josef Schwitter and Urs Heer, *Glarnerland: A Short Portrait* (Verlag Baeschlin, 2000), 29.

13. Interview with Roland Müller, in Stöckli and Müller, *Fritz Zwicky, An Extraordinary Astrophysicist,* 106.

14. This and subsequent quotes in the chapter are from an interview in 2015, in her home in Mettmenstetten, Switzerland.

15. Letter to Rösli Streiff, December 16, 1947.

16. "Germans Grateful for Library," *Los Angeles Examiner,* November 26, 1961.

17. The foundation was named for Johann Heinrich Pestalozzi, a Swiss writer and thinker born in 1746. He was a staunch advocate of universal education. His belief in the dignity of the individual foreshadowed Zwicky's own belief that every person was a potential genius.

18. F. Zwicky, "Free World Agents of Democracy," *Engineering and Science* (Caltech, November 1949), 12.

19. H. C. Honegger letter to Zwicky, June 29, 1954.

20. Or, in the words of Professor Virginia Trimble, who once was invited to his home for a memorable dinner, "The micro-Zwicky would be an average person. He was the whole 1,000 times rougher Zwicky."

21. F. Zwicky, *Morphological Astronomy* (Springer-Verlag, 1957), 11.

22. K. Rudnicki, "Philosophical Foundations of Zwicky's Morphological Approach," in *Morphological Cosmology* (Springer-Verlag, 1989), 418–426.

23. Rudnicki, "Philosophical Foundations of Zwicky's Morphological Approach," 6.

24. These quotes are all from F. Zwicky, *Everyone a Genius,* English version, unpublished, 1–29, Zwicky Archive of the Fritz Zwicky Foundation, in the Landesbiblioteck, the main library in Glarus, Switzerland.

25. "Jet Propulsion: The U.S. Is Behind," *Fortune* magazine, 141, undated article, Zwicky archive, Glarus.

26. Theodore von Kármán, with Lee Edson, *The Wind and Beyond* (Little, Brown and Co., 1967), 262.

27. William S. Barton, "Key to Producing Super-Mind Found," *Los Angeles Times* (March 26, 1950), 2.

28. "Machine to Travel through Earth Now Possible, Says Expert," *Eureka Times,* November 12, 1948.

29. Letter from Donald Loughridge, senior scientific advisor to the Department of the Army, December 28, 1948.

12. Bridges in Space

1. Official dedication of the Palomar Observatory, June 3, 1948. Remarks by James R. Page, chair of the Board of Trustees, California Institute of Technology, Zwicky Archive of the Fritz Zwicky Foundation, in the Landesbiblioteck, the main library in Glarus, Switzerland.

2. From a printed copy of the dedication in the Zwicky archive, Glarus.

3. Claude Stanush, "Big Eye Turns with the Earth," *Life* magazine (October 9, 1950), 103.

4. Allan Sandage, *Centennial History of the Carnegie Institution of Washington, Volume I, The Mount Wilson Observatory* (Cambridge University Press, 2005), 529.

5. The Gran Telescopio Canarias, however, like the Kecks in Hawaii, is made up of hexagonal segments, each about 1.5 meters across (59 inches). A modern equivalent to Palomar is the 8.2-meter (320-inch) mirror of the Very Large Telescope in Chile's Atacama Desert.

6. Tom Gehrels, *On the Glassy Sea, an Astronomer's Journey* (American Institute of Physics, 1988), 83.

7. *100 Days in Photographs: Pivotal Events that Changed the World* (National Geographic Books, 2007), 169.

8. David O. Woodbury, "Behold the Universe," *Collier's* (May 7, 1949), 20.

9. At least for the prime focus. The 200-inch also has a cassegrain and a coudé focus, for different work.

10. Woodbury, "Behold the Universe," 22.

11. Comment by James Gunn, Emeritus Eugene Higgins Professor of Astrophysical Sciences, Princeton University, December 11, 2017.

12. Zwicky was heard to boast about his eyesight, saying he was the only person who could keep a faint, 20th magnitude star on the slit of the 200-inch spectrograph in seeing 5, the sharpest possible image. Comment by Virginia Trimble, February 20, 2018.

13. John K. Lagemann, "The Men of Palomar," *Collier's* (May 7, 1949), 25.

14. Lagemann, "The Men of Palomar," 66.

15. Stars around the size of the sun spend their lives converting hydrogen to helium in nuclear fusion reactions in the interior of the star. More massive stars, eight solar masses and larger, convert helium to carbon and oxygen, on up to iron. Supernova explosions—both types—expel iron and heavier elements into the cosmos, essentially seeding the universe. There are some other production mechanisms, including something called the r-process, which the LIGO gravitational wave experiment showed in August 2017 comes mostly from mergers of neutron stars. This discovery led to the award of the Nobel Prize for Physics, in 2017. Honored were Kip S. Thorne of Caltech and Rainer Weiss and Barry C. Barish, both of the LIGO/VIRGO Collaboration.

16. Donald E. Osterbrock, *Walter Baade, A Life in Astrophysics* (Princeton University Press, 2001), 173.

17. Sandage's estimate of the expansion was, after some back and forth, 75 kilometers per second per megaparsec, close to the accepted range today of 65–74.

18. Some of Zwicky's clusters had "a good deal" of what Trimble calls substructure, "and might well have been called clusters of clusters (superclusters)." Even earlier, the Shane-Wirtanen counts had shown that isolated galaxies were few. Most were found in clusters (comments from Trimble during the editing of the book over a period of months in 2018).

19. "Bridges in the Cosmos," *Life* magazine (November 10, 1952), 51.

20. Alar Toomre and Juri Toomre, "Violent Tides between Galaxies," *Scientific American* (December 1973), 39–48.

21. In the lambda cold dark matter model, the "sheets & filaments (are) made of dark matter as is most of the mass of the galaxies and their clusters." Trimble comment, February 20, 2018.

22. Letter from Colonel Jack O'Mara to Fritz Zwicky, May 8, 1961.

23. Gehrels, *On the Glassy Sea,* 84–85.

24. Gehrels, *On the Glassy Sea,* 84–85.

25. Paul Wild, "Fritz Zwicky," undated, 9.

26. F. Zwicky, *Morphological Astronomy* (Springer-Verlag, 1957), 5.

27. Diary entry, July 15, 1952.

28. E. B. White, *The New Yorker* (June 30, 1956), 36–40.

29. Zwicky announced this plan in the *Chicago Sun-Times,* on March 16, 1956. He told reporter Ralph Dighton it was the best way to solve humanity's biggest problem, overpopulation.

30. The popular quiz show, "The $64,000 Question," had premiered to high ratings a year earlier. Clearly, White was not impressed. The quiz show genre imploded a couple of years later when it was revealed that contestants were coached and given the answers to questions.

13. Domestic Life

1. Interview with Marshall Cohen, Caltech, 2015.
2. "Constant Uproar," *Time* magazine (March 14, 1938), 41.
3. Interview with Zwicky's Swiss biographer, Alfred Stöckli, Glarus, Switzerland, March 2015.
4. Letter to Fran, November 1, 1964.
5. Letter from J. S. Warfel, May 29, 1951.
6. Quotes are from my interviews with Fran in Mettmenstetten, winter 2015.
7. As exacting with chocolate as with everything else, he wrote after one Halloween that Margrit had purchased 72 chocolate bars for the night, but only gave out 51.
8. Interview with Gerald Wasserburg, Caltech, 2015.
9. Letter to Rösli Streiff, August 20, 1948.
10. Letter to Fran, October 21, 1962.
11. Letter to Fran, February 15, 1963.
12. Letter to Fran, April 29, 1969.

Did Dark Matter Rattle Its Chains?

1. Interview at Caltech.
2. Davide Castelvecchi, "Controversial Dark-Matter Claim Faces Ultimate Test," *Nature* 532 (April 5, 2016), 14–15.
3. The experiment was a collaboration of universities and government agencies, including Caltech, Fermilab, Stanford, University of California, Berkeley, the National Institute of Standards and Technology, the University of Colorado, Denver, and others.
4. Most helium on Earth comes from the decay of uranium, thorium, and other elements underground, released by oil wells along with natural gas.

14. McCarthy and Sputnik

1. Letter to Fritz Zwicky in Switzerland, August 9, 1954.
2. Zwicky diary entry, May 15, 1955, Zwicky Archive of the Fritz Zwicky Foundation, in the Landesbiblioteck, the main library in Glarus, Switzerland.
3. Zwicky diary entry, April 19, 1955, Zwicky archive, Glarus.
4. Remarks excerpted in F. Zwicky, *Morphology of Propulsive Power* (Society for Morphological Research, 1962), 191.
5. Drew Pearson, "Why Cal Tech Ace Can't Work on Moon, Merry Go Round," *San Francisco Chronicle* (December 9, 1957), 29.
6. "Science: Missed Swiss," *Time* magazine (July 11, 1955).
7. Von Kárnán letter to Wilson, July 18, 1955.
8. Letter from Dan A. Kimball to Thomas S. Gates, June 3, 1957.
9. Zwicky, *Morphology of Propulsive Power,* 322.
10. Operation Lone Wolf folder, dated December 15, 1956, Zwicky archive, Glarus.
11. Remarks in an undated document, Zwicky archive, Glarus.
12. Diary entry about Aerojet sales meeting, August 8, 1955, Zwicky archive, Glarus.

13. "Free Radical Research Spurred By New Three-Year Program," *Aviation Week* (December 17, 1956), 31.

14. Letter to Tadeusz Reichstein, October 4, 1956.

15. This occurred at the Sixth Astronautical Congress, held in Copenhagen, Denmark.

16. Zwicky diary entry, July 29, 1955.

17. At this time, Zwicky was busy getting his book, *Morphological Astronomy,* ready for publication. In it, he aired his rage against the scientific establishment, savaging American science's obsession with "the voodooism of accuracy." On page 26, he referred to the "defeated men in astronomy who make it a profession to sneer at every determination of a position, luminosity or wavelength which is not accurate to one thousandth of a second of arc, one hundredth of a magnitude or one thousandth of an Angstrom." The point of this, he said, was "to prevent imaginative men from making too many discoveries." At this point, he did not name names. Later, he would not hesitate.

18. Fred L. Whipple and J. Allen Hyneck, "Observations of Satellite I," *Scientific American* (December 1957), 39.

19. Lee A. DuBridge, "DuBridge of Caltech Asks U.S. Tune Up Its Mental Powers," *Pasadena Star-News,* January 31, 1958, Zwicky archive, Glarus.

20. Whipple and Hyneck, "Observations of Satellite I," 42.

21. Graham Berry, "Rocket Expert Urges Space Research Aid," *Los Angeles Times,* October 14, 1957, 1.

22. "'Yes, No' Men Blamed for Science Lag," *Los Angeles Times,* November 6, 1957, 6.

23. "Rocket Expert Hits 'Quiz Show Mind,' Asks Science Contest," *Pasadena Independent,* December 9, 1957, 1.

24. Tape-recorded interview of von Kármán, owned by the Caltech archive, Pasadena, California.

25. F. Zwicky, "The First Shots into Interplanetary Space," *Engineering and Science,* Caltech (January 1958), 20–23.

26. Zwicky, *Morphology of Propulsive Power,* 183.

27. Zwicky, *Morphology of Propulsive Power,* 178.

28. It has been much debated in the years since whether Zwicky's pellets actually did leave Earth's embrace. But the *Los Angeles Times* proclaimed it a fact, declaring, "U.S. Fires Meteors into Space around Sun" (Graham Berry, *Los Angeles Times,* November 23, 1957). As did Dubin.

29. "Science: Defending Meteors," *Time* magazine (December 2, 1957).

30. Transcript of Zwicky's remarks, broadcast over KWKW in Pasadena, February 16 and 23, 1958, Zwicky archive, Glarus.

31. Transcript of Zwicky's KWKW broadcast remarks, Zwicky archive, Glarus.

15. Blue Stars and Quasars

1. Letter from W. Zisch to F. Zwicky, September 25, 1961.

2. Letter from F. Zwicky to W. Zisch, October 19, 1961.

3. Letter from F. Zwicky to W. Zisch, January 19, 1962.

4. Letter from A. Haley to Zwicky, March 20, 1962.

5. Undated note from Zwicky to von Kármán.

6. Letter from F. Zwicky to family, September 3, 1963.

7. Letter to Margritli and Franziska, September 3, 1963.

8. Alfred Stöckli and Roland Müller, *Fritz Zwicky, An Extraordinary Astrophysicist,* trans. Ian Gordon (Cambridge Scientific Publishers, 2011), 163.

9. Gagarin's size (he was only 5 feet 2 inches tall) was the reason he was chosen for the flight. His tart command, "Poyekhali!" (meaning "let's go!") became a watchword for the Soviet space program.

10. Letter from Zwicky to his daughter, Franziska, October 6, 1963.

11. Letter from Zwicky to his daughter, Franziska, November 19, 1963.

12. Phone interview, 2018.

13. F. Zwicky, "Compact Galaxies," *Acta Astronomica* 14, no. 3 (1964), 151–155.

14. Letter from D. W. N. Stibbs to Zwicky, May 29, 1971.

15. The earliest suggestion that so-called active galaxies with bright nuclei contain black holes at their centers is credited to E. E. Salpeter as well as to Ya. B. Zel'dovich and I. D. Novikov (1964).

16. The Harvard astronomer Robert Kirshner noted in a July 2018 email that it has been cited 724 times. "That's a lot," he wrote. "And people are still using it (16 citations in 2018 and 2017!)." Fritz Zwicky, *Catalogue of Galaxies and of Clusters of Galaxies* (Pasadena: California Institute of Technology, 1961–1968).

17. Letter from W. J. Luyten to Zwicky, February 5, 1959.

18. Letter to Margrit Zwicky, February 13, 1974.

19. Letter from W. J. Luyten to Zwicky, October 28, 1959.

20. Letter from Zwicky to W. J. Luyten, November 2, 1959.

21. But he was no delicate flower, either, being known in Minnesota for a gruff humor, which may be what drew him to Zwicky, who could carry gruffness and scorn to celestial heights. Luyten produced papers with titles like "Messiahs of the Missing Mass," "More Bedtime Stories from Lick," and the "Weistrop Watergate," about a young aspiring astronomer's PhD dissertation.

22. Steven Sotur and Neil deGrasse Tyson, eds., *Astronomy at the Cutting Edge* (New Press, 2000).

23. The National Academy bestows the medal. The letter to Luyten was dated April 17, 1964. Luyten's acknowledgement to Zwicky had come two days earlier.

24. Shirley K. Cohen, Oral History project at Caltech. Interviews conducted in April and May 1996. http://resolver.caltech.edu/CaltechOH:OH_Schmidt, 5.

25. Interview with Schmidt in his office at Caltech, May 16, 2014.

26. It's interesting how a person's walk can influence our impressions. In another incarnation, I was a journalist in Washington, DC, and had the opportunity to cover portions of the 1987 summit between President Reagan and the Soviet leader, Mikhail Gorbachev. One of my strongest memories was watching Gorbachev strolling the Rose Garden with Reagan, and being struck by his lumbering walk, very much the way Maarten Schmidt described Zwicky's.

27. 3C stands for Third Cambridge Catalogue.

28. 3C 273 turns out to be just about the brightest-known object of its kind. Today, according to Robert Kirshner, people can get spectra down to the 24th magnitude.

29. Oral History, Maarten Schmidt, Caltech, 27.

30. R. A. Sunyaev and N. I. Shakura (1973) contributed important work on the characteristics of the accretion disc that surrounds a black hole.

31. "Exploring the Edge of the Universe," *Time* magazine (March 11, 1966).

32. According to that theory, new stars and galaxies were continuously being created, while older ones became diluted by the expansion of the universe. It also suggested that most galaxies should be younger than the Milky Way.

33. In 2006, NASA released images from the Hubble Deep Field survey that showed hundreds of galaxies formed less than a billion years after the Big Bang. Those galaxies were smaller and blazed with birthing stars, evidence that galaxies formed from smaller pieces, like the dwarf galaxies Zwicky studied and believed were so important.

34. Another convincing piece of evidence was found two years after Schmidt's discovery, when Arno A. Penzias and Robert W. Wilson of Bell Laboratories discovered the Cosmic Microwave Background radiation left over from the Big Bang.

16. Banishment

1. K. I. Kellermann, "The Discovery of Quasars and Its Aftermath," *Journal of Astronomical History and Heritage* 17, no. 3 (2014), 267–282.

2. Allan Sandage, "The Existence of a Major New Constituent of the Universe: The Quasi-Stellar Galaxies," *Astrophysical Journal* 141 (1965): 1560–1578.

3. Kellermann, "The Discovery of Quasars and Its Aftermath," 227.

4. Zwicky letter to W. J. Luyten, June 7, 1965.

5. Zwicky letter to W. J. Luyten, June 7, 1965.

6. Fritz Zwicky and Margrit Zwicky, *Catalogue of Selected Compact Galaxies and of Post-Eruptive Galaxies* (F. Zwicky, Gumligen, Switzerland, 1971).

7. Kellermann, "The Discovery of Quasars and Its Aftermath," 277.

8. Phone interview of James Gunn, 2018.

9. Interview of Richard Ellis at Caltech, April 2014.

10. Not in all areas, certainly. Sandage's religious views were fairly unconventional for a scientist of his era. "If there is no God," he wrote, nothing makes sense. Writing at a time when the Big Bang was still a novel concept, he said the discovery of the "creation event does put astronomical cosmology close to the type of medieval natural theology that attempted to find God by identifying the first cause." But for Sandage, the discovery of the creation event wasn't yet finding God. Astronomers may have found the first effect, not the first cause.

11. Interview with Maarten Schmidt in his office, 2014.

12. Interview in her office, January 9, 2015, University of California, Irvine.

13. Interview of Gerald Wasserburg at Caltech, April 2014.

14. Zwicky letter to H. Babcock, August 4, 1965.

15. H. Babcock letter to Zwicky, November 19, 1965.

16. Interview with James Gunn, 2018.

17. Phone interview with James Gunn, April 28, 2017.

18. Phone interview with K. Thorne, February 7, 2018.

19. Zwicky Archive of the Fritz Zwicky Foundation, in the Landesbiblioteck, the main library in Glarus, Switzerland.

20. O. J. Eggen and A. Sandage, "Examination of the Evidence for the Existence of Pygmy Stars," *Astrophysical Journal* 148 (1967): 911.

21. George Getze, "Burrowing Jet Engines Called Vital for Future Moon Colony," *Los Angeles Times* (February 18, 1966), Part II, 1.

22. Letter to Franziska, February 18, 1972.

23. The survey was underwritten by the National Geographic Society. The largest percentage were taken by George O. Abell, who used them for his admired catalogue of galaxy clusters and planetary nebulae. Abell also wrote a wonderful introductory textbook for astronomy students.

24. Laurence A. Marschall, *The Supernova Story* (Princeton University Press, 1994), 110.

25. Letters to the Editor, "Reader Protests Smog Tax Plan," *The Independent* (December 19, 1960).

26. Lucie Lowery, "Astrophysicist Zwicky Gives WAGS Support," *The Independent,* Pasadena (November 22, 1965).

27. Letter to Margritli, September 12, 1966.

28. Letter to Franziska, June 21, 1967.

29. M. Zwicky letter to Margritli and Franziska, October 22, 1966.

30. Letter circulated by Zwicky to Caltech faculty, calling for direct election of Observatory Committee, June 1967. In an attached note, Zwicky writes, "Dr. Arp says this is too strong."

31. Letter to Franziska, February 4, 1967.

32. M. Zwicky letter to Margritli, February 17, 1967.

33. Letter from Milton Humason, April 25, 1967.

34. Jerry Lochbaum, "Fritz Zwicky: He Seeks Extremes of Universe," *San Antonio EXPRESS/NEWS,* April 15, 1967.

35. "February 1968: The Discovery of Pulsars Announced," *APS News* 15, no. 2 (2006).

17. Young America Going to Pot

1. Letter to Margritli and Franziska, September 18, 1962.

2. When they turned eighteen, each girl was given a gold ring with the family coat of arms.

3. Letter to Franziska, March 6, 1965.

4. Margrit was 36 at the time.

5. Letter to Franziska, April 14, 1968.

6. Letter from Stefan Staneff, June 1966.

7. Letter to Stefan Staneff, July 28, 1966.

8. Letter from Staneff, August 5, 1966.

9. Letter from Margrit to her oldest daughter, March 26, 1968.

10. Sue Avery, "Women in Automotive Class Have Terminology Trouble," *Los Angeles Times,* March 22, 1968, Part II, 10.

11. Letter to Margritli and Franziska, April 29, 1969.

12. Letter to Margritli and Franziska, March 15, 1970.

13. Letter to Margritli and Franziska, May 20, 1969.

14. Letter from Margrit to her daughters, May 2, 1969.

15. Letter from Fritz to his oldest daughter, May 12, 1968.

16. Letter to Franziska, June 10, 1968.

17. Letter to Franziska, June 12, 1968.

18. Letter from B. Vorontsov, December 1, 1967.

19. B. Vorontsov letter to Zwicky, undated.

20. A dark redoubt referred to by other denizens as "steerage," and "the engine room."

21. Interview with Maarten Schmidt in his office, 2014.

22. Phone interview with R. Kirshner, March 6, 2018.

23. Robert P. Kirshner, *The Extravagant Universe* (Princeton University Press, 2002), 144.

24. Phone interview with Robert Kirshner, March 6, 2018.

25. Kirshner, *The Extravagant Universe,* 145.

26. Note from H. Babcock, March 14, 1969, Zwicky Archive of the Fritz Zwicky Foundation, in the Landesbiblioteck, the main library in Glarus, Switzerland.

18. Magnificent Desolation

1. Letter to Fran, June 12, 1969.

2. Letter to the editor, *Time* magazine, July 17, 1969.

3. Letter from Margritli to Fran, August 19, 1969.

4. Letter from Margrit to her daughters, May 2, 1969.

5. Letter to his daughters, April 29, 1969.

6. Letter from Margritli to Fran, August 6, 1969.

7. It was founded in 1410. The University of Bologna in Italy is the oldest, founded in 1088.

8. Undated letter from Zwicky to Eleanor Ellison, his assistant.

9. In one of his diaries, he wrote that since science has discovered only order, it is unlikely that even an almighty God could act contrary to the rules prevailing in the universe "when he feels like it." Alfred Stöckli and Roland Müller, *Fritz Zwicky, An Extraordinary Astrophysicist,* trans. Ian Gordon (Cambridge Scientific Publishers, 2011), 186.

10. Letter from Fran to Margritli, March 22, 1972.

11. Undated letter to Eleanor Ellison, assistant and Caltech librarian.

12. Cygnus X-1 is a stellar-mass black hole, which was created by the collapse of a massive star. At present, it is about fifteen times the mass of the sun. It was the subject of a famous bet between Kip Thorne and Stephen Hawking, who believed it would not ultimately prove to be a black hole. Thorne won.

13. F. Zwicky and Margrit Zwicky, *Catalogue of Selected Compact Galaxies and Post-Eruptive Galaxies* (F. Zwicky, Gümligen, Switzerland, 1971).

14. Zwicky and Zwicky, *Catalogue of Selected Compact Galaxies and of Post-Eruptive Galaxies.*

15. She was also an important shoulder to lean on for female students at Caltech, who were much in the minority. Comment from Virginia Trimble, the physics professor, to me in a note, 2018.

16. Letter to Helen Humason, Milton's widow, June 26, 1972.

17. Zwicky notebook, Zwicky Archive of the Fritz Zwicky Foundation, in the Landesbiblioteck, the main library in Glarus, Switzerland.

18. Letter from Margrit to Margritli, May 5, 1972.
19. Letter from Margrit to her daughter.
20. Letter to Fran, May 31, 1972.

Mystery Unsolved

1. Unless the skeptics are right and a different theory of gravity is needed from general relativity, which would be a stunning discovery in itself.

19. A Scientific Individual

1. Meeting of the Royal Astronomical Society, in the Scientific Societies Lecture Theater, Savile Row. Transcript of Hoyle's remarks, *The Observatory* 92 (June 1972): 77–78.
2. Interview with Stöckli in Glarus, Switzerland, March 2015.
3. Zwicky Archive of the Fritz Zwicky Foundation, in the Landesbiblioteck, the main library in Glarus, Switzerland.
4. Letter to Luyten, October 13, 1973.
5. Alfred Stöckli and Roland Müller, *Fritz Zwicky, An Extraordinary Astrophysicist,* trans. Ian Gordon (Cambridge Scientific Publishers, 2011), 182.
6. Note in the Zwicky archive, Glarus.
7. Letter to Luyten, October 13, 1973.
8. F. Zwicky, "Nuclear Goblins and Cosmic Gamma Ray Bursts," *Astrophysics and Space Sciences* 28 (1974): 111–114.
9. "Zwicky's Ghost and Nuclear Goblins," *New Scientist* (June 27, 1974), 746.
10. "Zwicky's Ghost and Nuclear Goblins," 746.
11. It is now believed that mergers of two neutron stars produce forty Earth masses of gold, a treasure trove undreamed of, if one can figure out a way to get at it. (Note to the author from James Gunn, December 11, 2017).
12. Letter from Margrit to her oldest daughter, Margritli, February 8, 1974.
13. Note from Luyton, February 13, 1974.
14. According to Robert Kirshner, who was at the memorial, Greenstein told him privately afterward that he had cleaned up Zwicky's language out of respect for the occasion. He actually classified scientists as "eagles and shit-eaters," Greenstein said, according to this account. Greenstein's remarks are in the Zwicky archive, Glarus. The Kirshner remark from a note to me, 2018.
15. Zwicky archive, Glarus.
16. Zwicky archive, Glarus.
17. Bill Bryson, *A Short History of Nearly Everything* (Broadway Books, 2003), 31.
18. Letter from Margrit, June 6, 1974.
19. Letter from Margrit, November 1, 1976.
20. Letter from Margrit.
21. An instrument was finally named in his honor, the Zwicky Transient Facility, which saw first light in November 2017. Located at Palomar Mountain, with Shrinivas Kulkarni as principal investigator, the new camera can survey hundreds of thousands of stars in a single shot. It seeks out, fittingly, exploding stars, among other things.
22. Interview with Fran.

ACKNOWLEDGMENTS

At a writer's conference in Boston a few years ago, a visiting lecturer talked about a fascinating yet underappreciated character in science. He deserved to have his story told, the lecturer said, but he was so controversial that people had shied away. Having been a work-a-day journalist most of my adult life, I was intrigued. In journalism, after all, controversy is the coin of the realm. The more I learned about Fritz Zwicky, the predictions of dark matter and supernovae and neutron stars, the tectonic conflicts with the scientific hierarchy, his ultimate exile, the more I wanted to tell the story. The fact that this book exists, in spite of many rejections and impediments, is less a testament to the perseverance and insight of the author than to the fact that Zwicky is too important to be neglected. It's impossible to comprehend how the universe was stalked and ultimately tamed in the twentieth century without understanding the central role played by Fritz Zwicky.

I am grateful to Jeff Dean, executive editor at Harvard University Press, for seeing the potential in the story and for supporting me through the trials of bringing the book to fruition. For even though I had been a science writer at the *Los Angeles Times,* and though this book is aimed at the interested lay reader rather than the scholar and the scientist, I could not have done the job if not for the aid of experts and advisors who helped out.

Along with Jeff and his team at Harvard, I am grateful to two organizations that funded the project, allowing me to spend more than five years putting the book together. The Alicia Patterson Foundation gave me a grant that allowed me to travel to Switzerland to plumb the Fritz Zwicky Foundation archive. From that trip, I lugged home many hundreds of pages of copied diaries, scientific papers, letters, and newspaper and magazine clippings dating back many decades. Zwicky was a meticulous chronicler of his own life, and the volume of archival material is so large that my only regret was that I couldn't spend twice as much time there.

A grant from the Alfred P. Sloan Foundation allowed me to devote myself to the writing, as well as to hire experts in physics and astronomy to review the text for accuracy. I especially thank Doron Weber, vice president and program director at that foundation. This book, and many others, I suspect, would not exist without organizations like these, which have replaced traditional but fading sources of revenue for writers and journalists.

Among the experts who pulled me back, time and again, from stumbling into embarrassing errors, three are paramount: Virginia Trimble at the University of California, Irvine; James Gunn, professor emeritus at Princeton University; and Robert Kirshner at Harvard University. Virginia spent months coaching, educating, sometimes scolding, but always in the service of accuracy. Jim very generously read an early version and critiqued important chapters. Besides offering his expertise, Robert Kirshner provided colorful stories about Fritz Zwicky in retirement, as the old lion bent over his glass plates in a basement office at the California Institute of Technology.

Sunil Golwala at Caltech facilitated my trip in 2014 to the bottom of an iron mine in Minnesota, where scientists searched for the WIMP Miracle that would explain dark matter. Even though that miracle has not materialized, the trip helped me grasp the great efforts under way to unravel the mystery Zwicky posed eight decades ago. Thanks to Anthony Villano and Amy Roberts for explaining the workings of the Cryogenic Dark Matter Search experiment in that chilly hole.

In Switzerland, I am indebted to Max Weber for granting me unfettered access to the Zwicky archive in the basement of the Glarus town library. Thanks also to Alfred Stöckli, whose valuable biography, written with Roland Müller and published in English as *Fritz Zwicky: An Extraordinary Astrophysicist,* provided much of what I learned about the student Zwicky. The Zwicky Foundation generously supplied me with photos and illustrations, as did Caltech.

I am greatly indebted to two of Fritz's daughters, Franziska Zwicky-Pfenninger and Margrit (Margritli) Zwicky, for their memories and for sharing many personal letters revealing the family man. In addition I thank Margritli for translating a number of important letters from Fritz's Swiss-German.

I also owe a debt to friends and former colleagues at the *Los Angeles Times,* keen readers who offered smart advice about the writing. Chief among them are Jim Ricci, Deborah Schoch, and Dyanne Asimov. Jim and Deborah were more generous with their time and insight than I had any reason to expect. I also thank Rosie Mestel for her encouragement and friendship.

Any remaining errors of omission or commission belong solely to me.

Last but far from least, thanks to my agent, Bob Mecoy, who kept pitching when it seemed no one cared about a scientist most editors had never heard of.

ILLUSTRATION CREDITS

Illustrations on pp. 5, 21, 27, 32, 43, 54, 85, 95, 131, 151, 157, 169, 172, 179, 197, 206, 235, and 309 are courtesy of the Fritz Zwicky Foundation Archive, Glarus, Switzerland. Illustrations on pp. 66, 75, 87, 187, 250, and 257 are courtesy of the Archives, California Institute of Technology.

INDEX